ELECTRICAL Estimating Handbook

IRVING M. COHEN

A Construction Publishing Company Book

VNR VAN NOSTRAND REINHOLD COMPANY
NEW YORK CINCINNATI ATLANTA DALLAS SAN FRANCISCO
LONDON TORONTO MELBOURNE

Van Nostrand Reinhold Company Regional Offices:
New York Cincinnati Atlanta Dallas San Francisco

Van Nostrand Reinhold Company International Offices:
London Toronto Melbourne

Library of Congress Catalog Card Number: 77-20254
ISBN: 0-442-12152-0; formerly carried as ISBN 0-913634-24-7

Manufactured in the United States of America

Published by Van Nostrand Reinhold Company
135 West 50th Street, New York, N. Y. 10020

Published simultaneously in Canada by Van Nostrand Reinhold Ltd.

15 14 13 12 11 10 9 8 7 6 5 4 3

Library of Congress Cataloging in Publication Data

Cohen, Irving M
Electrical estimating handbook.

"A Construction Publishing Company book."
1. Electric engineering–Estimates. I. Title.
TK435.C63 1977 621.3 77-20254
ISBN 0-442-12152-0

PREFACE

This book is written for three groups of people: those who *do* electrical estimating, those who want to *learn* to do electrical estimating, and those who want to *understand* electrical estimates prepared by others.

For those who do electrical estimates, this book offers useful data, as well as guidance toward a more efficient and effective estimating technique.

For those who want to learn how to do electrical estimating, this book teaches a step-by-step method for making a detailed "take-off" (survey of needed parts and materials), for evaluating labor associated with the items in the take-off, and for calculating the overall cost. To give the beginner confidence in his estimating technique, the book furnishes a "ballpark" checking method.

For those who want to understand electrical estimates, the book gives labor values for common electrical items. In particular, for those who must judge the merits of change orders, the book provides a method of evaluating alternatives—for example, conduit quantities compared to wire, or vice-versa.

To obtain maximum benefit from this book, an electrical estimator should have a general knowledge of construction techniques for buildings ranging from wooden frame types to steel skyscrapers. He should be especially well versed in the type of building in which his firm specializes. He should have a sound knowledge of the National Electric Code and local codes and laws. He should understand power circuitry, control circuitry, and security circuitry as well as communication systems in general.

In addition to these prerequisites, it is desirable—although not essential—for an estimator to have an electrical engineering degree, a registered professional engineer's license, and an electrician's license. The professional engineer's license is particularly useful because it allows the estimator to design and estimate his own work and gives him the theoretical knowledge he needs to evaluate the designs of others.

Perhaps the most valuable attribute for an estimator is a background of several years in the electrical construction trade. Experience such as this develops the ability to visualize the true working conditions on a construction project and translate them into a realistic estimate of labor cost.

CONTENTS

SPECIFICATIONS AND DRAWINGS

Specifications, in conjunction with working drawings, designate the kinds, qualities, and methods of work and materials to be used in the construction of a building, as well as the responsibilities for the work and materials. Specifications and drawings are part of the contract by which the winning bidder will be bound legally. Therefore it is absolutely essential that a bidder understand them thoroughly to avoid oversights and unintentional omissions or mistakes.

Divisions

Usually, specifications are divided into 16 separate sections, according to the format approved by the Construction Standards Institute. Division 16, *Electrical*, is naturally the section of greatest interest to the electrical estimating profession. But this does not mean that other sections can be ignored. Quite the contrary, other parts of the specification document contain many provisions that affect electrical construction. The wise estimator will search thoroughly, from the opening line of Division 1, *General Conditions* (usually with *Technical* or *Supplementary Conditions* added), to the final period in Division 16, to determine the electrical contractor's responsibilities and liabilities.

Under *General Conditions*, for example, the estimator can determine whether the electrical contractor or the general contractor is responsible for temporary lighting and temporary power for equipment at the construction site. If an electrical contractor is responsible, the estimator must further determine whether the winning electrical contractor will automatically be given the job, or whether the general contractor will have the responsibility of supplying the temporary electrical facilities.

This same section of the specifications tells how much temporary lighting (in foot candles) is needed, who must maintain it, and who must pay the electric bill. It even covers responsibilities for such details as lamp-holders, bulbs, power cords, and outlets for hand tools. Such items can add up to a cost that will eat into the electrical contractor's profit if they are not taken into account in the estimate.

Other small but important and costly details covered in the *General Conditions* of the specifications include: (1) progress reports that the electrical contractor must turn in, (2) samples that he must furnish for examination or testing, (3) shop drawings (of lighting fixtures, panels, switchboards, etc.) he must provide. The dates for starting and completing electrical construction will also be noted, along with any bonuses or penalties for early or late completion.

As the estimator reads the General Conditions (and other sections too), questions are likely to arise. If the estimator cannot resolve a question himself, a call to the architect for clarification is in order. If the estimator decides that a questionable portion of the specification is not his responsibility, he should note his decision on his work sheet, referring explicitly to the appropriate page and line of the specifications. This careful record-keeping can be helpful if any contention arises at a later date.

Division 2, *Site Work*, tells the electrical contractor whether he is responsible for such things as temporary lighting, night security lighting, or temporary power while the site is being cleared.

It may come as a surprise that Division 3, *Concrete*, is also important to the electrical contractor. Here, the structural engineer tells the electrical contractor where conduits may or may not be run—in the slab, beams, or walls. The structural engineer usually spells out

TABLE I ABBREVIATIONS USED IN THIS BOOK

Al RC	Aluminum rigid conduit
BX	Armored metallic cable
Cu	Copper
EMT	Electric metallic tubing
FBO	Furnished by others
HOA	Hand-off automatic interlock
HP	Horsepower
HVAC	Heating, ventilating, and air conditioning
HW	Hard wall
KVA	Kilovolt-amperes
LP	Lighting panel
LVDP	Low voltage distribution panel
MCM	Thousand circular mils
MS	Motor switch; main switchboard
NEC	National Electric Code
ø	Phase
PVC	Polyvinyl chloride
RC	Rigid conduit
RGC	Rigid galvanized conduit
T	Thermoplastic
THHN	Heat-resistant thermoplastic
THW	Moisture and heat-resistant thermoplastic
TW	Moisture-resistant thermoplastic
XHHN	Crosslinked synthetic polymer nylon
XHHW	Weatherproof crosslinked synthetic polymer

the maximum size of conduit that can be allowed and where it can be placed in the pour. These and other restrictions and allowances in the concrete section bear an important relationship to the final bid price. Therefore the electrical contractor should clearly understand the structural engineer's intentions. Again, if the estimator has any doubts after studying the specifications, he should call the architect for clarification. A decision based on fuzzy understanding can come back to haunt the successful bidder to the tune of many dollars during construction.

Even though Divisions 4 *Masonry*, 5 *Metals*, 6 *Wood and Plastics*, and 7 *Thermal and Moisture Protection* seem unrelated to electrical work, they can still contain valuable information. There might be restrictions on membrane penetration, for example, that dictate roundabout conduit runs instead of direct ones. In addition to specific dos and don'ts, these sections help the electrical estimator learn something about other specialties—knowledge that can be a valuable aid in examining the working drawings. The same advice applies to Divisions 8 *Doors and Windows*, 9 *Finishes*, 10 *Specialties*, 11 *Equipment*, and 12 *Furnishings*. For example, suppose the section on doors and windows calls for a door-intrusion alarm system based on "normally open" magnetic switches. Is the electrical contractor responsible for cutting holes in the doors so that the magnets can be installed? Who should cut the door butt so the switch can be installed? Who is responsible for aligning the two parts? How critical is the alignment? What if it is discovered that the system does not work because the steel door dissipates the magnet's field? Who, then, is responsible for removing, relocating, and replacing the parts?

The section on equipment is especially pertinent to the electrical estimator, since much of the equipment is electrically operated. Here the electrical estimator can discover whether the equipment is portable or fixed, prewired or unassembled. He can discover, too, how much of the equipment installation the electrical contractor is expected to perform.

Divisions 13 *Special Construction* and 14 *Conveying Systems* often contain traps for the unwary. In these sections the estimator should again look for restrictions on placing conduit and cable. Who is responsible for fixtures or lighting within cabinets? Does the swimming pool come supplied with fixtures, or must the electrical contractor furnish and install them? Are fixtures up to electrical code standards? Are they prewired?

Division 15 *Mechanical Systems* usually covers plumbing, heating, and cooling. This section demands absolute attention because almost everything mechanical is electically powered. It is replete with potential problems for the electrical contractor because responsibility is often spread—sometimes arbitrarily—among several contractors. The specifications might state, for example, that "the HVAC (heating, ventilating, and air conditioning) contractor shall supply, install, and control-wire the baseboard heating units; the power wiring shall be done by the electrical contractor." Thus, even though the control wiring is electrical, it is not the electrical contractor's responsibility according to the contract document. The electrical contractor's sole responsibility is to bring power wiring to the units—and to make sure that they meet the electrical code, since the local inspector will blame the electrical contractor if they did not.

If duct heaters are called for, a jurisdictional question could come up. Should the sheet metal worker (who installs the heaters) or the electrician (who wires them) deliver the duct heaters to their proper floor and location?

Are pumps (supplied and installed by the plumbing contractor, but wired by the electrical contractor) listed by the Underwriters Laboratories? Could the local inspector fault the electrical contractor for wiring equipment that is not U.L.-approved?

If the electrical estimator has read the previous divisions with careful attention, he must now study Division 16, *Electrical*, with utter alertness. On his work sheet he will write the various bits of information that he will later use on the take-off sheets: such items as the minimum size of conduit allowed, the types of conduit or cable acceptable to the architect and engineer, the types of insulation required, the kind of conductors and their minimum size.

Now, the estimator is ready to back-check to make sure nothing has been overlooked. He quickly reviews the specifications, division by division, to ensure that all pertinent information has been included in his notes. He scrutinizes the electrical section even to the extent of comparing its headings with the table of contents and assuring himself that page numbers are in sequence and all pages are accounted for.

Drawings

The specifications should list the drawings associated with each of the divisions. It is important that the electrical estimator have a complete set of drawings. From the structural drawings, for instance, he can find the thickness of poured-concrete floor slabs and the height and width of columns. From the architectural drawings, he can determine the room finishes, the locations of partitions, the depth of hung ceilings, the thickness of insulation, the names of rooms and sections. From the plumbing drawings, he can find the locations and sizes of equipment to be connected to the electrical system.

The HVAC drawings should be compared to the specifications to ensure that the wording agrees. For example, if the baseboard heating drawings carry the legend "Furnished by the HVAC contractor, installed by the HVAC, power- and control-wired by the electrical contractor," the estimator should refer to the specifications for confirmation.

The HVAC drawings should be examined for disconnect switches, since the National Electric Code (NEC) states that each permanently mounted piece of equipment shall have a disconnect switch, usually within sight of the equipment that it disconnects. Disconnect switches are often omitted in drawings, but are necessary if the job is to pass inspection. The electrical estimator should decide which contractor—HVAC or electrical—is responsible for supplying and installing the disconnect switches. If the drawings or specifications do not assign the responsibility explicitly, a call to the engineer is in order.

The electrical drawings should be counted and their identification numbers checked to make sure the set is complete. The estimator should scan them slowly to get a feel for the size of the job. He should attempt to visualize the electrical construction, mentally routing feeder conduit horizontally and vertically from a switchboard to a panel, from an annuniciator panel to a thermal device, from an emergency generator to a panel. Visualization is not always easy because electrical drawings—unlike the other working drawings—are more a schematic plan view than a pictorial representation. There is no way of determining from an electrical drawing the height at which the various fixtures should be installed on a wall. An electrical drawing will show a receptacle, a switch, and an exit light all in the same position. In reality, these items are separated vertically; the specifications may indicate that the receptacle must be 18 inches above the finished floor, the switch 54 inches above the floor, and the exit light 8 feet above. The estimator, therefore, must visualize the electrical system in three dimensions.

The estimator can refer to the architect's section and elevation drawings to find routes by which the electrical feeder lines can pass to their destinations with a minimum of interference.One architectural drawing that can be troublesome is the reflected ceiling plan.

In the reflected ceiling plan, the architect shows how he expects the completed ceiling — including lighting and other appurtenances — to appear. The trick to understanding a reflected ceiling plan is to remember that — unlike other plans — it is not a bird's-eye view but a worm's-eye view. The estimator must imagine that he is lying on his back looking up at the ceiling when he examines this type of drawing.

The electrical estimator should coordinate his routes as much as possible with those used by the HVAC ducts and the plumber's piping. Formal coordination takes place after the bids have been awarded, but the electrical estimator would do well to anticipate the routes that the other contractors will use.

Organization of Electrical Drawings

The first sheet in a set of electrical drawings is usually the "legend" sheet (see drawing E1-1). This sheet depicts the symbols the designer has used and explains them. The legend sheet also contains notes in which the designer lists items that are required but not represented on the drawings by a symbol. In addition, the legend sheet contains any statements the designer wants to make about installation techniques or about nonstandard items he requires.

Usually, there is no table of contents for the electrical drawings. However, if the set of drawings is complex or if it has schedules, schematics, and details crowded on single sheets, it is advisable for the estimator to make his own table of contents. This practice will give him immediate access to information.

The next group of sheets is the lighting plans. These sheets show the location of lighting fixtures, their switches, wiring, and usually, panels. The home runs are not run back to the panels on the drawings. Instead, the line representing a home run is terminated in an arrowhead, usually with slashes to denote the number of wires in the conduit and a code for the destination (for example, LP3W 4,5 means that the conduit goes to Lighting Panel 3W, circuits 4 and 5).

The next group of sheets is called the power drawings. Actually, these sheets are plans for all electrical items that do not fall into the lighting category: receptacles, security equipment, intercommunication equipment, call equipment, power panels, motors for fans and pumps, electrical heating or melting equipment, and so forth. If there is a large number of such items, their plans are sometimes shown on separate sheets.

Power riser diagrams are on the next group of sheets. Sometimes called "one liners," these sheets show how the power is distributed, from the entrance switch to the smallest panel. Switches, panels, conduit, circuit breakers, fuses, and transformers for each level or floor are indicated. The emergency generator, the emergency transfer switch, and disconnect switches or circuit breakers for the elevators are also covered. The power riser drawings are schematics only; the exact location of each item on a level must be found by coordinating with the power drawing floor plan.

The power riser drawings specify the conductor metal, conductor size, conduit size, number of conductors within each conduit, the insulation, and—if conduit will not be used—the type of outside protection. For fuses, type and size are given. For circuit breakers, type, frame, size, and AIC (asymmetrical interrupting capacity) are noted. For transformers, primary and secondary voltages and the size in KVA are indicated. Similar information is furnished for panels and switches.

Switchboards and distribution panels on the power riser drawings are usually identified by a location code compiled by the designer. For instance, LP1W means Lighting Panel, Level 1, West Side. A panel whose designation starts with D is a distribution panel. LVDP means a low voltage distribution panel.

The next group of electrical drawings consists of the communication system riser diagrams. The title of each drawing defines which system it covers. There may be several systems on a sheet, or, if the job is large enough, each system will be drawn on a separate sheet. A riser drawing indicates the levels over which the system extends. It indicates the types of devices required, the room or area locations, and the number of wires between items. Sizes of conduit and wire are indicated, as are the number of wires in the individual conduit. A change in size is usually noted at the point where the change occurs. A diagram may use special symbols to depict equipment that is peculiar to a certain system.

The estimator should look for addenda on the drawings. An addendum is an addition to the drawings or specifications that is issued after the project has been offered to bidders. An addendum is usually indicated by a heavy outline with the number of the addendum enclosed in a triangle:

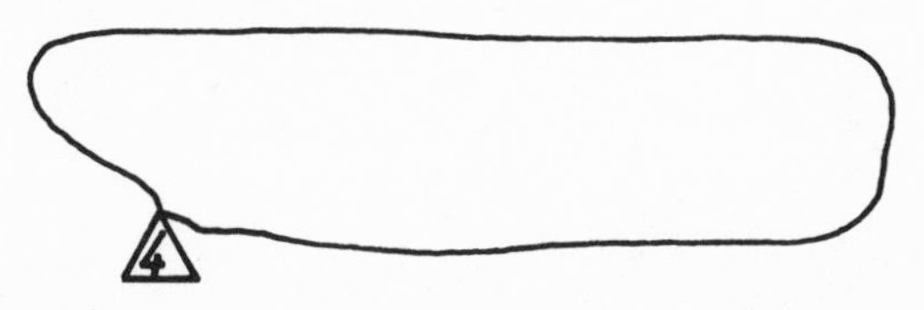

The number in the triangle is keyed to the addenda list in the title block of the drawing. (The Federal Goverment refers to addenda as "amendments.")

One caution: level lines or floors on riser diagrams should be checked carefully. Sometimes a designer adds dashed lines to denote that a riser continues into a space that normally belongs to another level. It is a good practice for the estimator to trace such lines with a red pencil so that he does not mistake their real level when he takes off the conduit runs.

The next sheets are the panel schedules, which tabulate the information that the riser sheets show diagramatically. A panel schedule is a list of panelboards that gives such facts as mounting, size of mains, distribution panels fed, location, and quantity and type of circuit breakers or fuses for each panelboard. (If size of conduit and wire and number of wires were not indicated on the riser diagrams, they will be listed here, usually in the "Remarks" column.)

Any special features of a panel are usually noted on the schedule sheet—if, for example, a panel is controlled by a remote contactor or a time clock, or has a "dump" type of shunt trip button for emergencies, the fact would be noted on the schedule.

A small addendum triangle is appended if a panel is covered by an addendum.

Motor control centers are usually included in the panel schedules. The destinations, conduit and wire sizes, and other pertinent information are given.

The final set of sheets contains details that the designer was not able to present on the previous sheets, and about which he wants to make his intentions clear. Among these sheets the fixture schedules can be found (although they are sometimes combined with the panel schedules). The fixture schedules list all the fixtures with a detailed description, manufacturer, and catalogue number. For public work,three manufacturers are usually listed for each item. If there is to be a deviation from the manufacturer's standard catalogue fixture, this fact is noted in a "remarks" or "notes" column.

Before the estimator moves on to the actual take-off, it is a good idea for him to review the HVAC and plumbing drawings. These drawings contain schedules that show the ratings of the equipment in kilowatts or horsepower. These schedules should be checked against the motor sizes shown on the power drawings. If the nameplate ratings are noted in kilowatts or horsepower, they must be translated to amperes for proper wire and conduit sizing.

THE ESTIMATING PROCEDURE

The estimating procedure should follow a natural sequence of steps that will familiarize the estimator with the site, the building, and special architectural features. The general strategy is first to take off—that is, count—the various fixtures, switches, receptacles, motors, and other electrical items. This is followed by take-offs of the conduit, cable, and wire needed to connect them to the electrical system. With the take-offs complete, the estimator adds up the quantities of the various items and calculates the amount of labor involved in installing each. He calculates the dollar value of the quantities and labor. Finally, the estimator considers factors such as weather and productivity to arrive at the bid price.

The individual steps are as follows:

1. Take-off of lighting fixtures.
2. Take-off of devices such as receptacles, switches, thermostats, fire alarms.
3. Take-off of motors and their controls, such as contactors, interlocks, switches.
4. Take-off of special equipment, such as for heating, lightning protection, air analyzing.
5. Take-off of conduit and wire for power and service risers.
6. Take-off of conduit and wire for motor branch circuits.
7. Take-off of conduit and wire for communication system risers. Includes such systems as fire alarm, security, intercom, supervisory control.
8. Take-off of conduit and wire for branch circuits.
9. Take-off of switchboards and panels.
10. Transfer of totals for the above take-offs to "price and labor" sheets. Determination of dollar value for materials and calculation of hour values for labor to install.
11. Transfer to extended prices and labor value totals to summary sheet. Determination of the final price in dollars to be bid.

Each of these steps will be considered in detail in the following chapters.

Aids for the Estimator

For take-off sheets, 17 x 11 inch accountant's work sheets ruled into 33 columns are best. For price and labor sheets, 8½ x 11 inch accountant's sheets ruled into four columns serve the purpose, although specially printed forms are available. Although the printed forms are good, the author prefers buff accountant's sheets for readability.

Counting the numerous fixtures, receptacles, and similar items is easier, faster, and more accurate if an automatic mechanical counter is used.

A rotometer is indispensable for measuring conduit, cable, and wire runs.

An adding machine—or better yet, an electronic calculator—is almost essential for handling the many additions and multiplications.

The estimator also needs a ruler, a magnifying glass (for examining drawings in detail), and red or carmine pencils (for checking off items on drawings as he takes them off).

No estimating office should be without a large collection of up-to-date catalogues—not just lighting fixture catalogues, but for any item that might be required in an electrical construction job. Besides giving price and discount information, catalogues provide valuable technical and installation information.

LIGHTING FIXTURES

The object of the lighting fixture take-off is to determine the quantity required of each type of lighting fixture in the building. To do this, the estimator goes over each lighting drawing, counting each type of lighting fixture one by one and entering the quantity counted on a take-off sheet.

The column headings—A, B, C, etc.—that the estimator writes on the take-off sheet are obtained from the lighting fixture schedule. (A sample take-off sheet is shown in Table TO-1; a sample lighting drawing appears in Drawing E2-1; a sample lighting schedule is in Table II.) Each heading represents a specific fixture, selected by the engineer. The estimator uses as many columns as necessary to list all the fixture types. More than one take-off sheet is used, if necessary, but the sheets are numbered in sequence to prevent confusion.

In the extreme left-hand column, the estimator lists the various lighting drawings he uses, such as E2-1, E2-3, E2-5, as he prepares to take off that sheet. He is now ready to go over each drawing, count the number of each type of fixture, and write the total in the appropriate column. To keep a record of his progress, the estimator puts a red check mark on the fixture symbol on the drawing as he counts it, usually in the upper right-hand corner of the square or circle representing the fixture. He does not mark through the letter, since this slows down or reduces the accuracy of backchecking.

It is possible to count fixtures mentally, but it is much easier and more reliable to use a mechanical counter (this applies to other take-offs too—not just to the lighting fixtures). The simplest such device is a counter that fits in the palm of the hand and has a ring which can be slipped on a finger. As the estimator checks off a symbol on the drawing with one hand, he can simultaneously move the finger on the other hand to register the fixture on the counter.

With a "clicker" of this type, the estimator can go over a lighting drawing and take off all the "A" fixtures, clicking and marking as he proceeds. When he has counted all the A's, he reads the amount on the clicker and writes it in the A column on the line for the drawing he has been using. He then resets the clicker and counts the B's. And so on. When he has counted all the fixtures on the drawing, he moves on to the next drawing, repeats the procedure, and continues until the fixture takeoff has been completed.

This single-type method yields the quantity of each type of fixture

TABLE II LIGHTING FIXTURE SCHEDULE

16510 LIGHTING FIXTURES

A. The Electrical Contractor shall furnish lighting fixtures complete with standard or special mounting frames, lamps, ballasts and other devices as required for a first-class installation. All fluorescent fixtures shall be furnished with ETL/CBM approved high power factor quiet operating ballasts to be for operation on 277 volts.

B. All ballasts to be Class "P" with internally protected automatic resetting thermal cutouts. Cutout to operate at 110° C ± 5%. All lighting fixtures shall bear UL approval with Class "P" ballasts.

C. All fluorescent lamps shall be rated for 12,000 hour life span and shall be cool white.

D. Suspended fluorescent fixtures shall be furnished with the proper pendant (as per manufacturers recommendations), swivel hangers and the proper canopy as approved by the Engineer.

E. All recessed fixtures shall be furnished with the proper mounting flanged for the type of suspended ceiling installed.

F. Where fixtures are recessed in fire rated ceilings, it is the responsibility of the Ceiling Contractor to furnish additional fire rated material above the fixtures.

G. All fixtures shall be supported from the masonry and steel building construction. Recessed fixtures shall be supported independantly with 2 #10 galvanized steel wires.

H. Fixtures shall be one of the following types:

"A" Sylvania PV 1021008 w/2-F96T12/HO lamps 20% uplight.

Westinghouse H2100 HSCS with 2-F96T12/HO lamps 20% uplight.

Curtis Electro PV252105 with 2-F96T12/HO lamps 25% uplight.

"B" Same as (A) only 4 feet long.

"C" Sylvania PVC 11008 with one F96T12/HO lamp

Westinghouse ST 196H with F96T12/HO lamp.

Curtis Electro BHO-196 with one F96T12/HO lamp.

16510.1

"D" Same as (C) only 4 feet long.

"E" McPhilben 15B-125 with 175 watt E-28 MOG. H37-5KC/C/E color improved Mercuryxopoc lamp. 277 volt.

"F" Sylvania #ET-10-BD-2404 with 2-40W-RS lamps and patterned acrylic lens. 277 volt.

Westinghouse #MG-240-14-TOP with 2-40W-RS and patterned acrylic lens. 277 volt.

Curtis Electro #BTF80-240 with 2-40W-RS lamps and patterned acrylic lens. 277 volt.

Note: Verify ceiling type prior to ordering recessed fixtures.

"G" McPhilben 15A-115 with 100W H38-4MP/C Color improved lamp. 277 volt.

"H" Sylvania #PVC 2604 with 2-60 watt, 800 MA. HO lamps.
Westinghouse. 277 volt.

"I" Curtis Electro.
McPhilben.

"J" Metallic Arts or equal type 50 ceiling mount.

"JW" Metallic Arts or equal type 3005 wall mount.

"K" Sylvania #6111-52-142, or equal, with 2-F40 lamps, and acrylic primatic lens patten 12 (277 volt).

"L" McPhilben #54-64-K-64 or equal, recessed exit sign with blank red lens panel and 3-40 watt A-19 lamps (120 volt).

"M" Spero #22005 (or equal) with 150 watt if lamp. 120 volt.

"N" See detail drawing 6 drawing E3-2.

I. All incandescent lamps shall be inside frosted long life extended service.

16510.2

(66) **SPECIFICATIONS SUB-SECTION 16510 — LIGHTING FIXTURES SUB-PARAGRAPH H.**

.4 Page 16510.1 Sub-Paragraph H. Revise the following fixture types to read:

Type "A" - Change the Sylvania Catalogue Number to PV-25-21008.

Type "J" - Change the Metallic Arts Catalogue Number to Type 5005.

Type "K" - Change the fixture to: Sylvania #AL-12-BD-2404 with 2-40 watt lamps. 277 volt or equal.

Type "L" - Change the fixture to: McPhilben #54-64, or equal, recessed exit sign with blank red lens panel and 2-25 T 6½ lamps.

Type "M" - Fixture shall be Crouse-Hinds Catalogue Number ARB 33 or equal W/150 watt I.F. lamp.

Type "P" fixture shall be McPhilben Catalogue Number 15BP-125N3 and 10 H hinged base. 175 E-28 mercury base or equal 277 volt.

Type "Q" fixture shall be Shalda #532 with 150 watt par 38 lamp, or equal 120 volt.

Type "R" fixture shall be McPhilben Catalogue #15BP-125 N3 or equal on a 10 H hinged base; 175 watt E-28 mercury lamp - 277 volt.

.5 Page 16510.2 Add the following Sub-Paragraph "J":

"J" All pendant mounted fixtures shall have a 8′6″ mounting height, from the finished floor, to the bottom of the fixture.

ADDENDUM NO.2

per drawing. The total of each type for all drawings is the important figure for bidding purposes, but tabulating on a per drawing basis makes it easier for the estimator to check the take-off later on. The estimator must not attempt to enter only a total for the entire set of drawings—he will have a difficult and confusing job of back-checking if he does. Instead, he enters the count in a page-by-page arrangement for a final total.

Besides the simple hand clicker, there are electronic and mechanical pencils which both mark and count. Counting with these devices employs the same principle as with the simple clicker: one type of fixture at a time is counted.

There are also more sophisticated counters that accumulate counts for several types of fixtures as they are counted. These multiregister counters have a separate button for each type of fixture. The estimator, however, must be careful to press the correct button. Although convenient, this type of counter is expensive.

The author personally prefers going over the entire sheet and taking off one type at a time, as opposed to a "block take-off." The one-type-at-a time method makes it less likely that the estimator will miss a fixture—as he goes over the sheet again and again, he may spot a fixture in an obscure corner that would otherwise be overlooked. And the repetitive scanning gives the estimator a better feel for the job. If the estimator has a multiregister unit, he should use it according to the single-type clicker method of take-off, but accumulate the A's, B's, etc., on the separate registers.

The estimator should beware of interruptions. If the phone rings or he is otherwise distracted, he should leave a distinctive marker at the last fixture counted so that he knows where to take up again when he returns. The marker could be a pencil pointing to the last fixture taken off, an eraser, or even a coin. If the take-off has been done horizontally—either right to left or vice-versa—the sheet should be checked for overlooked fixtures vertically.

DEVICES

"Devices" includes lighting switches, receptacles, and components of the fire alarm, security, intercom, and other systems—components such as speakers, fire alarm lights, horns, thermal detectors, pull stations, and telephones. The term as used here does not include motor switches or power control switches.

To take off the devices, the estimator must usually refer to both the lighting drawings and the power drawings. The lighting drawings usually show only lighting fixtures and their switches; the power drawings show the other devices.

To prepare a 17 x 11 inch work sheet for the device take-off, the estimator draws the device symbols (listed on the legend sheet, Drawing E1-1) as column headings, as shown in Table TO-2. He includes headings for the various lighting switches: S1, S2, etc. He does not concern himself with motor switches or power switches that appear on the power drawings; these will be taken care of later. If the estimator is using a multiregister counter, he inserts the symbols on the respective registers. For the device take-off, symbols are more important than words. But if there are closely similar symbols, it is advisable to distinguish them on work sheets and counters by adding a brief word description.

The estimator will ordinarily use two, three, or more work sheets for this take-off. He titles the sheets "Devices" and numbers them consecutively so that their sequence follows the lighting fixture work sheets.

The extreme left column is for the drawing number. As the estimator starts to take off a drawing, he writes its number in the left column to indicate that all take-off items on that horizontal line belong to that drawing.

The take-off technique is similar to that for the lighting fixtures. With a red pencil in one hand and a counter in the other, the estimator goes over a drawing, marking the right-hand or left-hand corner of the device's symbol and counting as he proceeds. As he finishes each item, he writes the count on the take-off sheet. When he has finished a drawing, he goes to the next one, writes its number on the take-off sheet, and proceeds in the same way.

Switches

The take-off usually starts with the lighting switches. The

estimator goes over each lighting drawing carefully, following the circuit lines as much as possible (See Drawing E2-1). He examines the doorways closely, since this is where most lighting switches are located. He looks for all types: single-pole, three-way, four-way, pilot-lighted, and others. Where switches are ganged, the estimator is careful to take off each one as a single switch, not in gangs.

No wiring is taken off. The time for this will come later.

A separate horizontal identity line is used for each drawing, even if there are only a few switches on a sheet. The time spent in making two entries is much less than the time that would be wasted in separating the counts for a back-check.

The experienced estimator need not take off all the single-pole switches first, then go back for the three-way switches, four-way switches, and so forth, one type at a time. Instead, because only a few different types are called for, he can take them off simultaneously, as they appear on a drawing. Even if the estimator works with a single clicker, he can train himself to count at least two types of switches at a time. For example, he can keep track of the single-pole switches as the first remembered acquisition and the three-way switches as the second. A count of 10, 4 would mean ten single-poles and four three-ways. He enters these counts under the proper symbols and starts counting the four-ways and pilot-lighted switches. The beginning estimator, however, should pursue a cautious course and count each type separately.

Receptacles

With the lighting switches counted, the estimator is ready to move on to the power drawings. The estimator uses the single-type approach. He starts with duplex receptacles, taking off only this type from the drawings. He moves along each circuit conduit, marking each duplex receptacle symbol in the upper right (or left) section with a red pencil as he counts it. (This method makes it easy to pick out a missed receptacle.) He does not trace the circuit with the pencil—this kind of marking is reserved for the conduit and wire take-off.

After counting all the duplex receptacles, the estimator goes back over the drawings and takes off the other receptacles, one type at a time: single, pilot-lighted, heavy-duty power, and whatever others there may be.

System Devices

System devices are the next items to be taken off. These devices usually appear on the power drawings (see Drawing E2-2), but are sometimes shown on separate drawings. The estimator starts with the fire alarm system. He counts the most numerous items first—usually the thermal detectors or manual pull stations. He ticks off the symbol and clicks the counter for each item as he did for the lighting fixtures.

The method is the same as before. In the left-hand column, the estimator writes the drawing number and under the appropriate symbol, he writes the number of items he has counted (see Table TO-2). He concentrates on one system at a time—in this case the fire alarm system—to make the take-off easier and more accurate. He continues until all fire alarm symbols on the drawing have been counted. He is especially watchful for one-of-a-kind items, such as the master fire station.

The estimator turns to the next power drawing and writes its number in the left-hand column. He repeats the take-off procedure. Even if a sheet has only one fire alarm item, he notes the sheet number, identifies it, and puts the item under its proper column heading. He continues through the drawings until he has counted all the devices in the fire alarm system.

Now, the estimator goes back to the first power drawing and takes off devices for the next system—perhaps the intercom system. He starts with the most numerous item on the drawing and continues with the others until the entire system has been taken off. As he turns to other drawings, he checks to see whether he has already listed them on the take-off sheet—listings should not be duplicated.

The estimator continues in this way until all systems, devices, and items have been taken off. At the end, the estimator may find that some symbols have no amounts in their columns. This is no cause for concern—the designer probably decided not to use those items. Or he may have used a standard set of symbols for the legend sheet, not all of which were called for. On the other hand, the estimator may find symbols on the drawings that do not appear on the legend sheet. Perhaps the designer put them in at the architect's request but forgot to include them on the legend sheet. Whatever the reason, they should be included on the take-off sheet, even if the estimator must write a description of them at the head of the column.

Checking the Work

When the estimator has completed the take-off, he goes back to the first sheet and scans it. Items that have been missed—those without a red tick in the upper right (or left) corner—will stand out prominently to his eye, which has been trained by the repetition. The estimator adds these unmarked symbols on each of the drawings to his totals. When he has done this, he will have double-checked his work quickly and accurately.

Incidentally, some estimators fill in the symbol with color pencil.

However, unlike placing a tick or check mark in the upper right corner (or, if the estimator prefers, the upper left), filling in the symbol does not produce a simple, repetitive pattern whose absence is easily spotted. The shapes of symbols are too varied for that. In addition, filling in the symbol is time-consuming and thus expensive.

SPECIAL EQUIPMENT

Special equipment consists primarily of electric heating equipment: thermostats, baseboard heaters, convectors, radiant heaters, in-duct heaters, and heaters imbedded in ceiling, floor, or wall. It also includes isolation transformers and lightning protection equipment. All of these items go on the special equipment take-off sheet. If there are a great many units, separate sheets for each type may be required.

Included in the special equipment needed for "Building 150," the building used for the sample take-off in this book, are snow melting mats on the outside steps. These appear in detail Drawing E3-3 and are taken off on Table TO-2.

Baseboard heating units can ordinarily be found on a schedule in the HVAC drawings, since the HVAC contractor is usually required to supply them. The estimator should take them off even though the electrical contractor is not responsible for supplying them—the various sizes of heaters have different wiring labor values that will have to be assessed later on. On the take-off sheet, the estimator lists the drawing numbers in the left-hand column. He lists the types as column headings. As he counts a heater, he ticks it off with a red pencil on the drawing.

Next, the estimator takes off the duct heaters. He makes column headings for the various sizes required, usually stated in kilowatts.

Ordinarily, duct heaters are supplied and installed by the HVAC contractor and wired by the electrical contractor. Sometimes, however, the HVAC contractor is assigned complete responsibility—including the wiring. The estimator should take off the duct heaters anyway, even though he will not include their labor values in his bid. Chances are that the HVAC contractor will ask the electrical contractor for a price. If the estimator has already done the take-off, he will be prepared to give a detailed answer. The general approach for the estimator is to take off all items, but to make sure that he knows his responsibilities and liabilities before including the items or their labor in the bid.

Many recent specifications contain a "Definitions" section that spells out the responsibility for supplying, installing, and wiring among the contractors.

Duct heaters often employ pneumatic thermostats. The electrical contractor, of course, is responsible for electrical thermostats only.

The estimator continues the take-off for the other special equipment. He classifies them according to size, length, wattage—whatever allows him to determine the purchasing price and labor to install.

For hospitals, isolation transformers often are required. They should be taken off on the special equipment sheet.

Lightning protection equipment should be taken off here, too. The lightning protection system will probably be supplied and installed by a "master label" electrician under contract to the electrical contractor. Nevertheless, the estimator should take off the equipment so that he can understand the bids and ensure that his subcontractor has installed according to the specifications.

The electrical heating equipment for "Building 150" consists chiefly of reheat coils and appears on the power drawings for the building, such as Drawing E2-2. The equipment is identified by drawing number and code (AC1A, HV1C, HV1B, etc.) on the take-off sheet, Table TO-3. Also listed on the take-off sheet are the power, volts, and phase of the heating units, and the disconnect switches required by electrical code (30F indicates a 30-ampere fused switch, for instance).

MOTORS

The motor take-off sheet is a complicated one. On it the estimator must list each motor along with its size (horsepower), voltage, phase, disconnect switch, fusing, starter, interlocks, indicators, and other controls. The headings for the columns therefore deserve special attention.

Column Headings

The extreme left column of the take-off sheet, as before, is for the drawing number. The column to the right of the drawing number is for the name of the equipment that the motor will drive—for example "Hot Water Heater" or "Air Compressor."

The next three columns are for these headings: "HP" (horsepower), "V" (voltage), and "Ø" (phase). (See Table TO-4.)

The column headings so far are straightforward, but what the estimator chooses for the remaining columns depends on the particular combination of motors in the building design. The headings should be selected so that each horizontal line contains all the necessary information pertaining to a given motor, and each vertical column contains information on the required quantities of items of a single type. The estimator will then be able to check quickly the individual items for each motor and to add easily the contents of any column to come up with the total number.

Before the estimator can decide on these additional headings, he must make a survey of the motors. He completes the columns for drawing number, equipment name, horsepower, voltage, and phase by referring to the power drawings. Other sources of data on motor size are the HVAC and plumbing drawings and the various schedules. When the estimator has tabulated this data, he will be able to determine what types of disconnect switches and other components are needed. He will also be able to organize the types on the take-off sheet.

For example, considering the motors listed, the estimator observes that the smallest motors are fractional-horsepower, 120-volt, single-phase motors and fractional-horsepower, 208-volt, single-phase motors. A simple motor toggle switch is all that is needed for the fractional horsepower motor, and so the estimator labels a column "1 Ø MS." The ½-horsepower, 208-volt, single-phase motor needs a fused disconnect switch according to the NEC. A 30-ampere capacity is adequate for a ½-HP motor. Therefore the estimator labels a column "1 ∅ Disconnect Switch, 30 F" under the general heading "208 V." (30 F indicates that the switch is fused and can handle 30 amperes.)

The estimator notices that the remaining motors operate at 480 volts, three phase, and that the largest of these is rated at 30 HP. A fused 100-ampere disconnect switch is appropriate for a motor of that size, and fused 30- and 60-ampere switches should be adequate for the smaller sizes. Accordingly, the estimator makes columns for "30 F," "60 F," and "100 F" under the general heading "480 V, 3 ∅ Disconnect Switch." Sometimes the specifications permit unfused disconnect switches; in such cases, the estimator would drop the "F" from the column heading. If the estimator is in doubt about the proper switch size for a given motor, he can consult the manufacturers' catalogues. Tables A-8 and A-9 in Appendix A ("Labor Values") can also be used to determine disconnect switch size.

Starters

The next group of columns is for magnetic starters. Beginning with the smallest motor and working up to the largest, the estimator assigns columns to the minimum size starters allowed by the specifications. For the motors in Table TO-4, the starter sizes range from 00 (for the small motors) to 4 (for the largest motor), and therefore the starter columns are labeled to include the sizes in between—00, 0, 1, 2, 3, and 4. Again, if in doubt about the proper size, the estimator can consult a catalog or Table A-10, Appendix A.

It is important to distinguish between the starters to be supplied and installed by the electrical contractor and those to be installed by the electrical contractor but furnished by another contractor. Accordingly, the estimator writes the general heading "By Electrical Contractor" over the starter columns just labeled. Now, he labels a similar set of columns and over them writes the general heading "FBO" (for "furnished by others"). The estimator need not be concerned with supplying the FBO starters, but must figure on mounting them and connecting power wiring to them—unless they are already part of the motor unit, in which case all the electrical contractor has to do is bring in power lines and connect them.

Remaining Columns

The remaining columns are for interlock and other control components. Depending on what is shown on the drawings or schedules, the estimator might make columns for "HOA" (for hand-off automatic interlock), pilot lights, limit switches, or alarms. A fan motor may require an interlock to trigger a fire alarm system, or to initiate a·certain sequence of operations as the fan starts or stops. Timers may be needed to operate damper motors. If an item is FBO, this should be noted in the column heading.

If a symbol is given on the legend sheet for a motor component, the estimator copies the symbol in the proper column on the take-off sheet. The symbol tells the estimator graphically what the designer wants to go with each motor and how he wants it controlled.

The final column should be labeled "?". Here the estimator can note unanswered questions or points that need clarification.

Total Take-Off

At this point, a few words in favor of "total take-off" (for which the take-off sheet organization just described is intended). Some estimators prefer individual item take-off, in which each motor circuit is taken off on a separate sheet. The disadvantage of the individual approach is that it leads to confusion and duplication of effort. The cost must be figured on a per-item basis, therefore prices of individual items must be looked up more than once if they are repeated in other areas. When the time comes to order parts, the total of similar items must be compiled from the separate work sheets—and the total must then be properly checked. With the total take-off sheet, these difficulties are avoided. The estimator can add the similar items merely by scanning the columns. He can check the results just as easily. And he need look up a price only once. And he can get a complete picture of the circuitry for each motor by scanning the horizontal lines.

Procedure

With the take-off sheet organized and the drawing numbers, equipment names, and motor characteristics tabulated, the estimator is ready to begin the actual take-off. He takes up the first power drawing and refers the first-listed equipment. The drawing will show the motor, its home-run designation, and its various controls. The estimator makes a mark in the appropriate column for each required item. He ticks off each item on the drawing as he lists it on the take-off sheet. (A tick mark means that he is finished with an item, so he does not mark it until he has gathered all the information he needs about it.)

The estimator reviews his notes or the specifications to determine whether an item should go under the FBO heading. As discussed in the chapter on specifications and drawings, responsibilities for supplying, mounting, and wiring equipment vary from job to job. Notes and references (such as "see note on H2-3") on the drawings are also helpful in this respect; such notes invariably designate the responsible contractor.

When he has taken off one equipment motor, the estimator moves on to the next. Again, he does not take off the wiring—he will come back to that later.

Sometimes information about the motor controls is not shown on a drawing. In such cases, there will be a note on the drawing to refer the estimator to a part of the specifications or to a separate schedule sheet or drawing for the information needed.

One final chore remains. The estimator adds up the various motors according to type and lists the totals along with the pertinent horsepower, voltage, and phase data as shown on the sample take-off sheet. This list tells the estimator motors to be connected—an important factor in calculating the labor involved in hooking up the motors to their power feeders. The list also tells the estimator the various voltages and wire sizes needed for these connections.

POWER AND SERVICE RISERS

This chapter deals with take-offs of the conduit and wire for the service and power risers. It will not concern itself with associated components such as switchboards, panels, fuses, and circuit breakers—they will be considered in a later chapter. The power risers are feeders—that part of the electrical system from which the branch circuits emanate. They are the circuit conductors between the service equipment (or the generator switchboard of an isolated plant) and the branch circuit overcurrent device. Service risers carry electric power from the point where it is furnished by the electric company to the main switchboards.

Power Risers

At the outset, the estimator reviews his notes for kinds of conduit and wire and limitations on them. Is the kind of conduit specified: electric metallic tubing (EMT), rigid galvanized conduit (RGC), aluminum rigid conduit (Al RC), or polyvinyl chloride (PVC)? Must the conduit carry a separate ground wire and, if so, must the ground wire be insulated? What is the minimum size of conduit required? What is the maximum conduit size permitted in poured concrete slabs?

The estimator takes the conductor requirements into consideration as well. Are the conductors to be copper or aluminum? Are they to be solid or stranded, or will both be used depending on size? What types of insulation? What sizes of cable? Any minimum size?

Setting Up the Take-Off Sheet

The left-hand column on the take-off sheet should be headed "Description" or "Items." Under this heading, the estimator writes the identity of the particular riser that he is taking off. For example, if he is using drawing E3-1 to take off the riser that runs from the Main Switchboard to Panel HP2, he writes "E3-1 MS to HP2" on the take-off sheet. (See Table TO-5 and Drawing E3-1 for samples of riser take-off sheet and diagram.)

For each of the types of conduit to be used, the estimator labels a column for each conduit size called for. Under "Rigid Galvanized

Conduit," for example, he would list ¾, 1, 1¼, 1½, 2, 2½, 3, 3½, 4, 5, and 6 inch conduit if these sizes were called for by the specifications or drawings.

The estimator does the same for the wires. For each type of insulation, he makes a column for each wire size called for. He reds out the runs on the riser drawings as he transfers the length of the run to the take-off sheet. He then knows that each has been included in the take-off.

Like the motor headings, the riser headings give the estimator a detailed description of the item horizontally and allows him to total vertically.

Measuring Conduit Runs

The take-off consists of measuring the lengths of the various conduits and wires and noting the measured values in the proper column. To do this, the estimator uses a rotometer. He will be working with several drawings at once, so it is necessary to have plenty of room in the desk area.

There is one drawing that the estimator will refer to constantly during the course of the take-off: the power riser sheet. If possible, the estimator should pin or staple this drawing over his desk. If his desk is in a corner, so much the better; he can pin the drawing on the wall at his right (or left). It should be firmly attached so that he can trace each run with a red pencil as he takes it off. On the wall directly in front of him he should attach the architectural elevation drawing or section. The other drawings should be in front of him on his desk, where he can leaf through them easily.

To illustrate the take-off technique, we will follow the path of a riser from its origin at the Main Switchboard to its destination at HP2, a power panel. The estimator first locates the Main Switchboard (MS) on the riser diagram (Drawing E3-1), which shows that this switchboard panel is on level 24 of the building. Next, he locates Panel HP2 on the riser diagram and finds that it is on level 36.

To find the location of these pieces of equipment on their respective floors, the estimator turns to the power diagrams. On the power diagram for level 24, he find that he must turn to Drawing E3-2, Detail 13, to locate the Main Switchboard in the switchboard room, between coordinates 4 and 5, south of the "J" line. On the power diagram for level 36, Drawing E2-4, he finds HP2 between coordinates F and G on line 6. The riser then, runs from level 24 coordinates 4-5 south of the J line to level 36 between coordinates F and G on line 6.

The estimator refers to the architectural drawings—particularly the vertical sections—to see what provision the architect has made for vertical conduit runs from level 24 to level 36. He finds that the architect has included a mechanical chase, running the entire height of the building, for electrical conduit, water pipes, and air ducts. (It is ordinarily a good idea, incidentally, for the estimator to make sure that there will be enough room for electrical conduit in such a chase by examining the mechanical drawings for the dimensions of the air ducts.)

The estimator studies the relative positions of the chase and the two pieces of equipment on their respective floors. He finds that Panel HP2 on level 36 is mounted on the chase wall. But Main Switchboard No. 1 on level 24 is a considerable distance from the chase. The architectural drawing for level 24, moreover, reveals obstacles to an easy overhead approach to the chase—the floor is divided into small rooms with ceilings containing many ducts.

The alternative is to approach the chase through the ground. The structural drawings reveal that the floors of the building will be poured concrete slabs 12 inches thick. The specifications (or the estimator's notes) indicate that it is permissible to put the required 3½-inch RGC conduit in the slab provided that: (1) the conduit does not cross over any other conduit, (2) other conduits are spaced at least 2 inches away, (3) the conduit is laid no closer than 2 inches to the top or bottom of the slab. The floor, or below the floor in the grade, therefore, is chosen as the medium for the level 24 approach to the chase. The estimator, however, must remember later in the take-off to keep other conduit the prescribed minimum distance away from it, if the conduit is run in the floor slab.

One point that sometimes causes confusion: when the estimator runs a conduit in a ceiling, he is actually running it in the floor of the story above. In other words, to get some of the information needed about the ceiling of the first floor, the estimator may have to refer to the structural drawing for the second floor.

The estimator is now ready to determine the length of the conduit. He takes the rotometer and moves it over the power drawing to measure the distance from Main Switchboard to the chase, through the floor of level 24. He does not measure the distance "as the crow flies;" instead, he moves the rotometer over a gentle arc (the arc allows extra conduit for possible variations in the building).

Now, the estimator looks at the elevation section drawing to find the vertical distance to level 36. The conduit will traverse only from level 24 to level 36 if it enters Panel HP2 through the bottom. But it must traverse additional height if it enters through the top or back. In either case, the estimator notes the number of vertical feet to reach level 36.

Finally, the estimator measures with the rotometer the short distance from the chase to HP2. To the three measurements—horizontal distance on level 24, vertical distance in the chase, and horizontal distance on level 36—the estimator adds the distances of the switchboard and panel above or below their floor or ceiling, as the case may be. The sum is the total length of the conduit. Suppose the sum is 140 feet. This value would be entered for "MS to HP2" in the 3½-inch column under the Rigid Galvanized Conduit heading.

The estimator should not take off the fittings, straps, connectors, couplings, terminations, or clips that normally accompany conduit. In the author's opinion, such parts do not represent enough value to warrant even the small amount of time spent in counting, marking them down, and carrying them in the totals. (It might be wise to count them in change orders, but not in competitive bids.) Instead, the estimator may, if he wishes, add a small distance—say 10 feet—to the run total to cover for the cost of small parts.

One item the estimator should count in the take-off, however, is conduit bends, or "sweeps." If the conduit diameter is 1¼ inch or larger, sweeps can amount to a significant cost and labor factor. For the 140-foot run of 3½-inch conduit just measured, the estimator needs six 90-degree sweeps (or simply "90's"). On the take-off sheet, the estimator marks the number of 90's to the left of a slash line next to the length of the conduit run, like this: 6/140.

Determining Wire Length

It remains to calculate the amount of wire to be installed in the conduit. The schedule sheet, Drawing E3-4, under the heading "MS" in the upper lefthand corner, tells the estimator that the conduit from Main Switchboard to HP2 carries eight wires consisting of No. 350 MCM copper in two separate 3½-inch rigid galvanized conduits. The length of each wire is the length of the conduit, plus a little more to allow for connecting at the switchboards and panels. Usually 10 feet per switchboard or panel is adequate unless the estimator knows (from the switchboard detail drawing) that the board is very large or the wire enters the board on the side opposite the side at which it is connected. The individual wire length is therefore 160 feet, or a total of 1280 feet for the four wires in each of the two conduits. The estimator enters 1280 under the 350 MCM, XHHW column for "MS to HP2."

The estimator has now completely taken off the riser from Main Switchboard to HP2. He traces over the riser on the drawing with the red pencil to show that he is finished with it.

Other Risers

In taking off the remaining risers, the estimator adheres to a pattern. In this regard, the riser from Main Switchboard to HP2 was selected for the take-off because it provided a good example. In practice the estimator would select the first riser on the right coming out of the switchboard. The take-off can thereby proceed from right to left (alternatively, the estimator could choose a riser on the left and proceed from left to right) in an orderly fashion. The estimator would then follow the next riser to its destination, read the conduit and compute the wire, and mark down the quantities in the proper columns and line. He would continue until he has redded in every riser coming out of the switchboard.

The riser drawings usually contain notes, and often, addenda. The estimator must consider these comments from the designer and, as he acts on them, mark them off with a slanting red line.

Ampacity

Specifications invariably contain a statement to the effect that "the contractors shall follow all local codes and the National Electric Code." This blanket statement protects the designer by making the electrical contractor responsible for all errors or omissions involving codes or local rulings.

Electric codes can seriously affect a riser take-off. When risers are to be laid in poured-concrete slab floors on grade, such slabs are often classified as wet or damp locations. This classification imposes severe limitations on ampacity—the maximum current that a given wire size may safely carry. For example, suppose that the riser from Main Switchboard to HP2 is required to supply 650 amperes per wire. If the slab through which this riser runs in level 24 is considered damp, the current is heavy for the wire, because 350 MCM copper, insulated with XHHW, is good for only 620 amperes. (If the riser were in a dry location, it would be good for 650 amperes.)

Regardless of whether they run through damp locations, it is good practice to review the conduit and wire sizes to make sure that they meet the code requirements. Unless he does so, the estimator cannot be sure that the architect and designer have ascertained that the sizes conform to the code. And no matter who is at fault, the electrical contractor will be held responsible for code violations or omissions.

The estimator should not forget that derating enters into the take-off. The NEC requires that wires be derated (their ampacity reduced) when there are more than three in a conduit. The reader is referred to Article 310 of the National Electrical Code.

Service Conduit and Wire

Service conduit and wire carry electric power from the point where it is furnished by the electric company to the main switchboards. The point of origin may be a transformer in a local vault (maintained and operated on the premises by the utility), a transformer pad off the premises, or a transformer on an overhead pole. The riser diagram will show the point of origin.

Service conduit and wire are taken off in much the same way as ordinary risers. It is important, however, to distinguish between items that will be supplied and installed by the utility and those for which the electrical contractor is responsible. If the entry is underground, who is responsible for trenching and backfilling? If entry is overhead, will the utility attach the service wires, or is the electrical contractor required to install poles and run conduit and wire? Will the utility charge the contractor for any work.

From the site drawings (see drawing E1-1), the estimator takes off the conduit and wire with a rotometer. He also takes off ground rods, clamps, cadwells, weatherheads, transformer lugs, special cabinets for meters, and fittings. He also takes off the ground bus that runs around the perimeter of the transformer vault or switchboard room. On the take-off sheet, the estimator identifies the drawings in the horizontal lines and notes the items in the vertical columns. He watches carefully for the note "Supplied by utility" and disregards any item so described, unless labor is required to install it.

Site Lighting

While the estimator examines the site plan for service conduit, he can also check it for site lighting and parking lot lighting. Site plans often carry outdoor lighting fixtures mounted on the building as well. The estimator makes a note of such lighting so that he can come back to it later. The fixture should be added to the lighting fixture take-off and, later, the conduit and wire should be added to the branch circuit take-off.

Clearing Up Gray Areas

If the estimator finds an error or ambiguity in a riser drawing, he must ask the electrical designer about it. First, however, following protocol, he calls the architect to get his permission—it is important to keep the architect informed. Often, the designer is not local and therefore does not share the estimator's knowledge of local rules. (A local code takes precedence over the national code. The NEC itself states that the "governmental body exercising legal authority" is the final judge.)

The designer will usually issue an addendum to clarify the situation. The estimator who raised the question will have the advantage of advance notice of the addendum.

The entire riser sheet should now have been redded out: risers, notes, addenda—every word or line that is part of the drawing or has any meaning related to the drawing. The riser take-off is now complete.

MOTOR BRANCH CIRCUITS

Although the wires that carry power to motors are usually called feeders, they are actually branch circuits, if they have fuses in the disconnect switch, since a branch circuit is defined as "that portion of the wiring system between the final overcurrent device protecting the circuit and the outlets." Regardless of the definition, it is convenient to take off the motor branch circuits on the power riser take-off sheet. This is because most motors require larger conductors—comparable to those used for risers. Motor wire size is usually larger than No. 10 gauge. The estimator concerns himself only with the conduit and wire for motor power—he does not take off control wiring at this point.

It is a good practice for the estimator, before he starts the take-off, to check the horsepower ratings of the motors shown on the HVAC and plumbing drawings and schedules against their wire sizes. The ampacity of the wire must be adequate for the motor horsepower, and include the 125 percent factor that the code calls for.

The take-off procedure is the same as that for the power risers. In the left-hand column of the riser take-off sheet, the estimator identifies the motor circuit to be taken off by its drawing number, source, and destination (see Table TO-5). Assume that the estimator is taking off the circuit running from HP2 to HV2B on the same level as HP2 at coordinates 8,G (Drawing E2-4).

Again, the estimator uses the technique of locating the items by their coordinates and visualizing the run between them. In order to do this in a thorough manner, however, the estimator must refer to other drawings if necessary. Since HV2B is in a storage space, the estimator refers to the plumbing drawings. He checks to see if a plumbing run or fire main might interfere with the electrical run—or vice versa. (The days when the first contractor to get in a space "owned" it are gone. Today's complexity demands that contractors cooperate and coordinate their efforts.) Now, the estimator turns to the HVAC drawings to look for duct runs that might interfere with the electrical run. He checks duct sizes and hanger positions to be doubly sure.

While he is looking at the HVAC drawings, the estimator can find out which side of the motor the terminal box is on so he can allow enough conduit and wire to reach it.

The estimator examines the architectural drawings to see if there are partitions that might obstruct a straight conduit run. It might be advisable to run horizontally in one area and drop directly into the disconnect switch.

It is wise to check whether there are other motors feeding from the switchboard in the same general direction as the HV2B feeder. The estimator may be able to save labor by installing several runs on the same hangers.

If the estimator plans to run the conduit in the floor slab, he refers to the structural drawings to find the restrictions imposed by the structural engineer.

The estimator is now ready to take-off the particular motor branch circuit. He runs the rotometer over the conduit path he has decided on. To the distance measured, he adds the vertical distances from HP2 and HV2B, and writes the total in the proper conduit column on the take-off sheet. To find the wire length, he adds to the conduit length an allowance of 5 feet for the motor connection and 10 feet for the switchboard connection, and multiplies the total by the number of wires. If the conduit run to HV2B is 70 feet, the wire required will be (70 + 5 + 10) x 4, or 340 feet in this case in the conduit and wire size noted in the schedule for Panel HP2 on Drawing E3-4.

Disconnect Switches

As mentioned earlier, the NEC requires a disconnect switch for most electrical equipment. The estimator has presumably counted all the motor disconnect switches on the motor take-off sheets.

Ordinarily, a motor disconnect switch with lockout capability on the switchboard will meet the requirements. However, if the motor is remote from the switchboard (as HV2B is, since it is quite a distance away from HP2), it is necessary to provide a local disconnect switch. In such cases, the estimator should consider the feed from the switch to the motor in the motor circuit take-off. Often, flexible metal conduit (such as Greenfield) is used. Or, if the motor drives a pump, a waterproof flexible conduit (such as Liquitite) may be specified. These types of conduit can be expensive. If they are used, the estimator should add columns for them—and whatever connectors go with them—on the motor circuit take-off sheet.

Fittings

The estimator will also include certain fittings in the motor circuit take-off. It may be necessary to feed a motor from a tee fitting or—for extra strength and rigidity—from a floor fitting. All the estimator need do is add columns to the motor circuit take-off sheet and count the items. Some judgement is needed, of course. If the value of the fittings is trivial, or if the estimator can allow for them more efficiently by adding a few extra feet to the conduit, the estimator should not bother to count the fittings.

Completing the Take-Off

To complete the take-off of the run to HV2B, the estimator goes back to the power drawing on which HV2B is located and reds out the home-run designation. He then proceeds to take off the other motor circuits, redding out with his pencil as he goes.

Relations with Other Contractors

As the estimator takes off the various motor circuits, he should keep in mind the cautions mentioned in the chapter on specifications and drawings. The answer to the question "Which contractor is responsible?" can be elusive. If there is doubt, it is essential that the estimator get a firm answer from the architect, engineer, or designer—a large sum of money could be involved.

The question of duplicate responsibility sometimes comes up. One section of the drawings or specifications may assign responsibility for an item to the electrical contractor while another section may assign it to the HVAC or plumbing contractor. The estimator obviously needs an authoritative answer to resolve such questions.

Incidentally, a nonelectrical contractor will frequently ask advice on ordering motors for which he is responsible. Whether or not such advice has been asked, the burden of ensuring that the motors meet the size, phase, and voltage specifications falls on the electrical contractor.

Testing

The electrical contractor is expected to participate in testing the electrical equipment—even equipment he has not supplied. Accordingly, the estimator should include a column labeled "Testing" on the motor circuit take-off sheet and enter a reasonable number of hours in it for each test required.

The electrical contractor usually has an electrician stand by during tests to identify switches and other parts of the system for the HVAC or plumbing contractor. Even if the equipment manufacturer sends a field engineer to run a test, an electrician must be kept at hand to change wires if necessary. The estimator can check the specifications to see how long the test is expected to run and how much set-up time is involved. He can then list the total time in the "Testing" column.

Some specifications require the electrical contractor to familiarize the customer's maintenance people with the electrical equipment. If the job requires such an educational period, the estimator should carry some hours for it.

Motor Control Circuitry

It is best to take off the control circuitry for the motors separately from the motor feeder branch circuits. There are several reasons for doing this. The conduit wire sizes are much smaller for the control circuits. Motors and their home runs are usually on the same level and reasonably close to each other, whereas the control circuitry can be located almost anywhere in the building.

The most important reason is that control circuitry is usually "scheduled." That is, control circuits are indicated on the power drawings by coded letters—the estimator must look up the letters in a schedule on the HVAC, plumbing, or other electrical drawings. A schedule might indicate, for example, that a motor must be interlocked with other motors; pumps; shutters; or pressure, limit, temperature, or flow switches. Working out routes for scheduled control circuitry requires referring to many different drawings. It often requires much thought on the part of the estimator, and can be time-consuming. The take-off should therefore be handled as a separate task from the motor branch circuit take-off.

The motor control circuit take-off is handled in much the same way as the power riser take-off. The estimator determines the various sizes and types of conduit needed, sets up columns for them on a "Motor Control Circuit" take-off sheet, visualizes the route for each circuit, and measures the route, taking extra caution in noting the number of conductors in each circuit. There might be as many as 12 conductors in a conduit of 150-foot length. This equals approximately 2000 feet of wire.

COMMUNICATION SYSTEM RISERS

Communication systems include fire alarm, coded call bells, burglar detection, central music, closed circuit television, buzzer entrance signal, nurses' call, paging, public address, telephone, and many other systems.

Communication systems are shown as riser diagrams because they usually spread through the entire building. Each communication system is a distinct entity, and often must be run in separate conduit, particularly if shielding is required. Depending on how large the job is, the systems will be shown on separate diagrams, or some or all of the systems will be combined on a single diagram.

In this chapter, we will take off a typical communication system riser (Drawing E3-1), using a fire detection system as an example. The estimator has already taken off on the device sheet the devices (such as thermal detectors) used in the system. Now he is ready to take off the conduit and wire.

The take-off is easier if two people do it. But this is not always possible, and so we will demonstrate the one-man take-off.

Fire Detection System

The estimator titles a work sheet "System Risers and Branches" and assigns the sheet its proper page number (see sample take-off sheet in Table TO-6). To start, the estimator examines the riser diagram to determine the types and sizes of conduit and wire, just as he did for the power riser take-off. He lists the sizes and types as headings on the take-off sheet. In the left-hand column, he notes the name of the system—in this case "Fire Alarm System."

The number of wires in each conduit is usually indicated by the number of slashes across the conduit line on the drawing.

Sometimes the wire and conduit sizes are not indicated on either the drawings or specifications. If the estimator can find no indication of size—even in the designer's notes on the system riser drawing—he should choose the minimum size allowed by the code. It is a good practice to keep a code book handy for this purpose.

The estimator can make a quick check of the earlier count of devices on the device take-off sheet by counting the devices on the riser diagram. If there is a difference in quantity, he should use the larger number. (It may seem easier and faster to count devices from the riser diagram instead of the way suggested in the chapter on devices. The latter method, however, gives the estimator greater familiarity with the system.)

The estimator needs—in addition to the system riser drawing—a legend sheet (to identify unfamiliar symbols), the power drawings, and the architectural drawings. The riser drawing shows, for each building level, which devices are used. It does not show, however, where on each level the devices are located. To obtain this information, the estimator needs a drawing that shows the device positions in relation to rooms. The power drawings usually contain this information, although it may be necessary to refer further to the architectural drawings to locate specific room numbers.

The estimator goes over the entire riser diagram carefully. He locates the master box, the annunciator panel, and the control panel. He acquaints himself with the arrangement of the fire detection system into zones. He makes sure that the designer has conformed to the local fire code.

The estimator begins on the lowest level with the zone on the far left (Zone 1 in Drawing E3-1). For the first device on the extreme left in this zone, the estimator reads off the room number or location marked on the riser drawing. He finds this location on the power drawing. (If the room number is not on the power drawing, he determines it from the architectural drawing and writes it on the power drawing in the proper place.) This location or room number will be the starting point for the take-off; we will refer to it as point 1. The estimator is starting at the extreme-left device on the system riser drawing. On sheet E3-1, the device to the left on the lowest or first level (level 24) is noted as a fire alarm manual station, directly below a "rate of rise" thermodetector operating in Zone 1 and located in Room G101. Also in Room G101 is another rate-of-rise detector, a manual pull station, and a fire alarm horn with an integral light. The estimator now turns to Drawing E2-2 and in Room G101 (Truck Court) notes the location of these devices and any others that are in the same zone. He measures the distance in ¾-inch conduit, as noted on the riser. He then "scoots" home to the fire alarm section of the control unit which he has located in the Security Room. The note on E2-2 directs the estimator to the Security Room layout detail. He will then progress to the devices on the right in the same manner. If he prefers, he may start from the right and work toward the left—the important thing is to move consistently in one direction.

On the system riser drawing, the estimator ticks off this first device. He moves to the next device to the right and continues as he did for zone 1. He reads its room number or location. He finds this point

(point 2) and marks it on the power drawing. He runs the rotometer between points 1 and 2 on the power drawing. He adds to the rotometer reading the "ups" and "downs"—the distances of the devices above the floor or below the ceiling, depending on which route he has chosen for the riser. The total—the rotometer reading plus ups and downs—is the length of conduit between the devices. (It is important to include the ups and downs. They affect the wire take-off as well as the conduit take-off and could throw the totals off by as much as a factor of 100.)

One convenient way of handling ups and downs without mental arithmetic is to draw lines to scale on the power drawing for the up and down distances. Then all the estimator has to do is run the rotometer over the up or down line to add it automatically to the conduit run.

The estimator now ticks off the second device on the riser diagram and moves on to the next device in line. He reads its room number or location and marks the point (point 3) on the power drawing, as before. He checks to see what size conduit links points 2 and 3 and what size and number of wires it carries. If the sizes and number are the same as in run 1-2, the estimator measures run 2-3 with the rotometer, letting the measurement accumulate with the 1-2 measurement. He includes the ups and downs, as before.

If run 2-3 is not additive—if it requires a size of conduit or wire or number of wires different from run 1-2—the estimator writes the total accumulated on the rotometer on the take-off sheet in the appropriate conduit column. (The estimator totals the ups and downs for the run if he has not been automatically including them on the rotometer.) He multiplies the total value for the run by the number of wires in the conduit (in this case six No. 14 wires) and writes the result in the appropriate wire column on the take-off sheet. (For an alternate method, see "A Short Cut" in the Branch Circuit chapter.)

The estimator takes off the lowest level and goes on the upper levels until he reaches the top. Then he takes off any runs that lead back to the control panel. He reds out the risers as he proceeds, as well as the devices and notes. When he comes to a change in conduit size, wire size, or number of wires, he writes down the accumulated riser length before going on. He pays particular attention to the junction boxes—this is where wires are usually gathered and routed together to the control box.

Just as with the power risers, the estimator must plan conduit routes carefully to avoid interference with ducts, pipes, and other conduit. For a conduit larger than 1¼ inch in diameter, he takes off the sweeps. He writes them on the take-off sheet in the same way as for power risers: 4/150 for four 190-degree sweeps in a 150-foot length.

At the control panel, the estimator looks for home runs to a power switchboard battery rack and to a municipal fire alarm system and takes them off. He consults the specifications to detemine the types of remote lamps (one filament or two) and other accessories connected to the control panel, and estimates the wire and conduit to be run for these components.

The fire system annunciator, if there is one, is often located a sizable distance away from the control panel. Because of the distance, and because it usually contains many wires in large-size conduit, it is important to include the annunciator run in the take-off.

The pull boxes and splicing boxes appear only on the system riser diagram (if they do not appear, the estimator should include those that are required by standard electrical practice). Junction boxes for the system risers are usually larger and deeper than normal. The estimator should make sure that there is enough room at the junction box to accommodate this larger size; if the space is cramped, installation time will be longer. The estimator should therefore make a note of such difficult-access locations so he can take them into account when he computes labor costs.

The designer's notes can tip the estimator off to some obscure requirements that might have a significant bearing on the estimate. For example, a note might refer the estimator to certain HVAC drawings for smoke detector locations. The estimator might thereby find that the electrical contractor is responsible for supplying and wiring smoke detectors—even though none appear on the fire alarm riser drawings. Another note might state that the electrical contractor is required to interlock the HVAC system with smoke detectors. Even though no such interlock wiring may be shown on the fire alarm risers, it is up to the estimator to allow for it in his bid, if he believes it is included.

Other Systems

The technique for taking off other communication system risers is the same as the one used here for the fire alarm system. Essentially, the level and room number for the devices, the size of the conduit, and the number of conductors are found from the system riser diagram, the device location is found on the power or architectural drawings, and the distance between devices is measured on the power drawing. The ups and downs are added.

Telephone System

The telephone system is something of a special case. Ordinarily, the electrical contractor is asked to install a sleeve for telephone wires during construction of the building. The telephone company itself will pull the wires. The estimator is therefore concerned only with routing and taking off the sleeves, or supplying and installing empty conduit, with or without pull wires, according to the specifications.

BRANCH CIRCUITS

A branch circuit is that portion of a wiring system between the final overcurrent device protecting a circuit and the outlets (lighting fixture, receptacles, fixed equipment, etc.). The procedure for taking off branch circuits differs from that for power risers in two important respects: (1) the branch wire and conduit are considerably smaller, the wire size seldom exceeding No. 6 gauge and the conduit rarely being larger than 1¼-inch diameter; (2) less precision is permissible with branch circuits because the value of wire and conduit is less.

The estimator sets up the take-off sheet along the same lines as a power riser sheet, making column headings for the various types and sizes of wire and conduit. He may also need columns for cable such as BX (armored metallic cable) or Greenfield, since these are often permitted for branch circuits. The lighting and power drawings note the conduit, wire, and cable information needed for the column headings. The left-hand column of the take-off sheet is for the identifying drawing numbers.

Lighting Branches

The estimator starts the take-off with the branch circuits for the lighting fixtures. On the lighting drawing (Drawing E2-1), he selects the last fixture in the circuit, preferably in the lower right or lower left corner (he will work from right to left or left to right, as he prefers). He traces with the rotometer the entire circuit connected to this fixture, stopping when he arrives at the fixture with the home run designation. (He will take off the home run later, unless the wire and conduit are the same size). He reds out the circuit with a pencil.

He repeats this procedure for all other lighting circuits having the same size conduit and wire and number of wires as the circuit he has just measured. When he has finished, he writes the total rotometer measurement in the appropriate conduit column on the take-off sheet (Table TO-7). It is not necessary to write down the measurement for individual runs on the take-off sheet. Instead, the estimator saves himself time and effort by letting the distances for like runs accumulate on the rotometer and writing down the total.

In tracing a circuit with the rotometer, the estimator should consider the nature of the building, since it will affect the route he chooses. If the building is of steel frame construction with hung ceilings, the route will be different from that in a wood frame or concrete building. In any event, the estimator does not follow a straight line or the circuitry on the drawing; he runs the rotometer in a gentle arc from fixture to fixture. This gives him enough extra footage to cover the cost of fittings, hangers, and bends. The estimator is watchful for stairwells and elevators; circuits must be routed around, not through, these architectural features.

Computing Wire Quantity

Assume that the estimator's first measurement is of ¾-inch rigid galvanized conduit runs containing two No. 12 THHN wires. His rotometer total is 500 feet and 100 fixtures were involved in the runs.

For 500 feet of conduit containing two wires, the wire length is 1000 feet. The estimator must also allow for wire used in crossing junction boxes and splicing in fixtures. The amount of this allowance should be liberal—about 2 feet per box or fixture is not excessive. For the 100 fixtures, therefore, the estimator needs 200 feet per wire, or 400 feet. Now the total wire is 1400 feet.

Now, the estimator goes through the same procedure for the other conduit-wire combinations. He multiplies the conduit length by the number of wires and adds an allowance of 2 feet per wire per fixture. The estimator enters the total values of ¾-inch RGC and No. 12 THHN for drawing E2-1 in the proper columns, as shown in Table TO-7.

The estimator does not bother to count wire nuts, connectors, tape, or other items needed for splicing at the fixtures. A small addition to the wire quantity (about ½ percent of the total wire) or, later on, to the labor charge will take care of these inexpensive but necessary items.

A Short Cut

At the beginning of this chapter it was stated that it is not necessary to be quite so precise in the branch circuit take-offs as in the other take-offs. The estimator can therefore save time by lumping together all conduits of the same size, even though they do not contain the same number of wires—provided that the wires are of the same size and insulation. This is done by estimating the percentage of two-wire, three-wire, four-wire, etc., runs in the total, then taking the appropriate percentage of the total conduit run. For example, suppose that in the 500 feet of ¾-inch RGC, roughly 50 percent contains two wires, 30 percent contains three wires, and 20 percent contains four wires.

The total wire length would be:

(50% of 500 ft) x 2 wires = 500 ft

(30% of 500 ft) x 3 wires = 450 ft
(20% of 500 ft) x 4 wires = 400 ft

This adds up to 1350 feet of THHN wire in the conduit.

To this total, the estimator must add 2 feet per fixture for splicing. Again, he estimates the portions of the fixtures devoted to two, three, four, etc., wires. He multiplies the number of fixtures by the percentage and the number of wires. For example, if 30 percent of the 100 fixtures are four-wire, the calculation would be (30% of 100) x 4 = 120 feet. The estimator would allow 120 feet of THHN wire for the four-wire fixtures, plus the number of feet found by similar calculations for the other fixtures.

The estimator can be even less precise if a conduit has a fairly even mix of two-wire and three-wire sections—roughly 50-50. He can regard it as all three-wire in his take-off without introducing any serious error.

The percentage method results in a slight overestimate of wire requirements. However, the amount of money involved is small, and worth the time saved. The estimator should be sure to adjust the percentages of the wire mix, however—they may change from drawing to drawing.

Home Runs

After the estimator has taken off the fixture runs, it is time for him to consider the home runs. These are easy to pick out on the drawing at this point because the other runs have been redded out.

The home runs are usually taken off separately from the fixture runs to avoid jumping from one conduit size to another in the calculations and rotometer measurements. The procedure, however, is essentially the same: the estimator visualizes the route for the home run and measures it with the rotometer, including, of course, the ups and downs.

Power Branches

Taking off the power branch circuits is somewhat more complicated than taking off the fixture branch circuits. The fixture circuits rarely exceed a 20-ampere rating, but the power circuits can have many different and higher ratings, depending on their use. The estimator must therefore deal with a larger variety of wires and conduits.

Larger jobs show the power branches on a separate drawing from the lighting branches, as in Drawings E2-1 and E2-2. In this case, the drawing numbers are sufficient to distinguish the power runs from the lighting runs on the take-off sheet (see Table TO-7). But if power and lighting branches are combined on the same drawing, the estimator should use separate take-off sheets, marking one "Power Branches" and the other "Lighting Branches." By making this distinction, the estimator makes it easier to back-check.

The receptacle circuits are taken off first. The estimator starts in one area, as he did with the lighting circuits, and follows the conduit or each receptacle circuit with the rotometer to its home run. He does not yet take off the home run unless, of course, it is the same size as the receptacle circuit conduit and wire.

The route of a conduit will not necessarily coincide with that shown on the branch circuit drawing; it is often necessary for the estimator to exercise ingenuity in planning a route. If the floor is a concrete slab, the estimator's job is somewhat simpler since he can route the conduit directly in a straight line from one receptacle to another in the slab. As always, however, he must look out for interference from other conduit in the slab.

The estimator allows for the distance of the receptacle above the floor. If the distance is 18 inches—the usual distance—he allows 3 feet for each receptacle (18 inches up and 18 inches down).

The estimator has a choice of making an exact take-off (measuring each conduit-wire combination separately) or using the percentage method described on page 41 to average out the number of wires in the conduit. The first method is considerably more time-consuming, and not that much more exact.

After the estimator has completed the receptacle circuits, redding them out as he goes along, he takes off the home runs. As always, he is careful to route them in a realistic way.

Other Circuits

In addition to the receptacle circuits, there are other types of power branch circuits to be considered. If the building will be heated by electrical baseboard units, the wiring for these will appear on the power branch circuit drawings. Even if the baseboard heaters are supplied by another contractor, the electrical contractor is ordinarily required to wire them and to make sure that they meet the specifications for size and electrical characteristics.

The take-off of the baseboard circuits is similar to that of the receptacle circuits. The estimator starts with one baseboard unit and follows the circuit to its home run. He repeats the procedure for all the circuits, then takes off the home runs. On the take-off sheet, he enters the data on a line identified by the drawing number and the word "Baseboard."

Other power branch circuits include circuits for small air conditioning tempering reheat coils. These are taken off in an identical fashion. Emergency circuits, which must be run separately in their own conduit, are taken off as above.

Other Types of Conduit

The preceding discussion has focused on a rigid conduit, or "pipe," type of installation. The take-off methods are equally applicable to all types of conduit and cable—subject, of course, to the peculiarities of the particular medium.

For example, if the estimator is taking off the surface metal raceway known as Wiremold, the runs on the wall must be parallel to the floor, with sharp 90-degree bends to the vertical. All fittings and elbows should be taken off so that both material and labor costs can be computed.

If underfloor ducts are used, the estimator should pay particular attention to underfloor junction boxes and fittings, again because of high material cost and labor values.

Sleeving is a necessary part of poured concrete construction. When sleeves are indicated (usually in notes, details, or schedules or located exactly in the structural drawings), they must be included in the estimate. The estimator lists the required sleeves by size and counts them on the take-off sheet if there is a large number of them. Otherwise, he allows for sleeves as a factor in the conduit take-off. Although supplied by the electrical contractor, they are ordinarily installed by the concrete contractor.

Expansion fittings are usually required every 100 feet in a conduit, at expansion joints in the concrete slab, and at tie-ins between buildings. The specifications will tell the estimator where they are needed and the drawings will show—usually on detail drawings—the types that are required. Expansion fittings must be counted individually; they are too expensive to "allow for." The estimator lists them on the power branch take-off sheet.

Cable

In areas where cable is allowed—in partitions and hung ceilings, for example—the estimator must draw on his experience in planning routes. In addition, labor is a big factor in cable installation, since many holes may have to be drilled through partitions. Therefore, if the labor value for cable installation (calculated from the hours per 100 feet of cable given in Appendix A) is to be accurate, the estimator must be particularly careful to make accurate measurements. He should also be sure to allow extra footage for splicing, and avoid the tendency to run "tight."

SWITCHBOARDS AND PANELBOARDS

Switchboards and panelboards are single panels or assemblies of panels containing switches, overcurrent protective devices, and buses. Switchboards often contain instruments and are often accessible from the rear as well as the front, depending on the type. Panelboards feed or control circuits of small or large capacity; they are usually designed to be placed in a cabinet or against a wall or partition and are therefore accessible only from the front.

Switchboards and panelboards are usually supplied by manufacturers as a package that includes transformers, circuit breakers, fuses, switchgear, rectifiers, current transformers, etc. The electrical contractor should take off all this equipment so that he can ensure that the manufacturer provides the correct equipment in the proper amount.

However, because the estimator cannot get the individual prices for the components of the package, the switchboard and panelboard take-off is the one take-off that can be transferred directly from the drawings to the pricing sheet. Pricing sheets will be considered in detail in the next chapter. An example of a switchboard and panelboard pricing sheet is shown in Table TO-8.

Circuit breakers and fuses merit special attention. The designer might require fuses of a new, special type, and a complete set of spares—a spare fuse for every fuse holder. The sizes of the fuses are also important because they affect the labor cost.

There are certain hidden costs that can become apparent during the switchboard and panelboard take-off. The estimator should make a note of them as they arise so that he can factor them in later on. For example:

- The electric company may supply power meters and current transformers, but the electrical contractor may be required to wire them.
- For large pieces of equipment, the electrical contractor may incur drayage or rigging charges.
- If jurisdictional questions come up, the electrical contractor may run into out-of-the-ordinary labor costs. The electric company may want its own people to install its transformers, for instance. If the electrical contractor's electricians would balk at this, the estimator will

have to allow for "standby" men who will assist the utility personnel.

The panelboard or switchboard schedule sheet provides the estimator with the data needed for the take-off (see schedules in Drawing E3-4). On it are listed all the service, distribution, and branch equipment. Also listed are the number of circuits for each panel, the size of the wires, and the size of the conduits. The estimator checks the schedule against the power riser drawings to ensure that all panels, switchboards, transformers, and other distribution equipment are included.

As a further check, the estimator reviews the drawings, identifying and locating each protective device or fuse. He ticks off each breaker or fuse on the drawings as he checks it against the take-off sheet. However, he does not tick off the items on the schedule sheet—this is done when the breakers or fuses are transferred to the pricing sheet. As a final check, the estimator ticks off the circuit breakers on the main distribution panel drawing.

The estimator examines the schedule sheet to see if any panels are contactor controlled (by a photoelectric cell, for instance) and whether the panel manufacturer or the electrical contractor is expected to provide the contactors. Manufacturers often do not regard contactors as part of the package and the item might therefore be missed.

The notes on the schedule sheet will reveal if the panels must be double-bused. If so, more labor and material is required.

The NEC allows only 42 poles in a branch circuit panel. The estimator should keep this restriction in mind as he adds the poles of the circuit breakers in a panel. If there are more than 42, a double panel is needed. For example, twenty-two 20-ampere single-pole breakers, six 20-ampere double-pole breakers, and six 30-ampere three-pole breakers add up to 52 poles. A double panel is needed, therefore. The same restriction applies to fuses; no more than 42 fuses are permitted per panel.

As the estimator gains experience, he will often be able to go to the pricing sheet with only a one-line notation on the take-off sheet for the panelboard or switchboard. He will be able to recognize repetitive sizes and combine them mentally. At the start, however, the estimator should not attempt to condense or abbreviate—he should list all the items on the take-off sheet.

PRICE SHEETS

The quantities on each of the take-off sheets must now be totaled and transferred to "material and price" sheets—or, simply, "price" sheets. On the price sheets, the material costs and labor time for all items will be entered.

Thus, on the sample take-off sheet TO-1, the estimator writes the word "Total" on the left, near the bottom of the sheet. He goes across the sheet, entering the sum of each column on the Total line. He does the same for all other take-off sheets in sequence.

If the estimator is working alone, no item totals should be transferred to the price sheet until all take-off sheet totals have been checked. However, if the estimator has an assistant, the checked item totals may be transferred earlier—but only when a given item is complete and checked by someone else for the entire job. Such items as devices, lighting fixtures, and systems equipment can usually be transferred in this way. Items that usually require further totalization before they can be transferred include conduit, wire, motors, and motor starters and switches. These should not be transferred until the subtotals of all the take-off sheets involved have been added.

As a total is transferred to a price sheet, it is redded out on the take-off sheet with a pencil. Even if a column is unused and therefore has no entry for the total, a red dash is drawn through it to indicate that it has not been overlooked.

Calculating Aids

The additions should be done with an adding machine that provides a printed tape of the computation. The printed tape allows the estimator to check his figures quickly, since the numbers can be compared to those in the columns on the take-off sheet.

Even more helpful than an electromechanical adding machine is an electronic calculator. A printed tape is not really necessary because an electronic calculator is so fast and easy to use that it is often preferable to re-add the figures instead of checking them. The memory of an electronic calculator is another big advantage. A running total can be kept of the "extended" items (unit costs or labor values multiplied by the quantity of items required for a job). When the last item has been extended, the user pushes the memory recall key to see the extension total. There is no need for him to add the numerous item extensions.

Regardless of the electronic or electromechanical aid employed, each column on each take-off sheet should be checked twice: once by the estimator and once by someone else.

Price Calculations

The author prefers to use standard accountant's work sheets with four-column ruling as price sheets. Besides these, there are pricing forms published by certain electrical associations, primarily the Minnesota Electrical Association (for sale to all) and the National Electrical Contractors' Association (for sale to members). Regardless of the form used, the technique is the same.

Tables TO-8 through TO-12 show representative price sheets (accountant's work sheets were used). For simplicity, the columns have been headed by the letters A through G. The meaning of these letters is as follows:

A. "Adder" or "more detail" column.
B. Quantity.
C. Description of item.
D. Cost per item.
E. Extension of item cost (B × D).
F. Labor value of the item.
G. Extension of the labor value (B × F).

The labor values in columns F and G are in decimal hours (1.00 hour represents 60 minutes, 0.25 hour represents 15 minutes, etc.) to simplify the arithmetic.

An example will illustrate the use of the pricing sheet. Consider a 20-ampere, 277-volt single-pole switch costing $2.12. (Prices are obtained from catalogues and wholesalers.) Assume that 70 such switches are required. Under column A, the estimator writes the letter "S," indicating that the unit is a switch box that the electrician can install from the floor. (Other possible entries in column A are "C" for a ceiling box, for which the electrician would need a ladder, and "B" for a box installed in a wall bracket, for which the electrician would need mechanical aids. Each of these designations denotes a different labor value.) These "adders" are totaled separately and entered under C (ceiling), B (bracket), or S (switchbox) at the end of each section.

Under column B, the estimator writes the quantity, 70 (see Table TO-9, line 25). Under column C, he writes "S1" for single pole switch. Under column D, he writes the unit cost, $2.12. Under column E, he writes the extended cost for the units, 70 × $2.42, or $148, rounded off to the nearest dollar. Under column F, he writes .24, a value he obtains from a chart of labor values; the value represents the labor time in hours for installation of this type of switch.* Under column G, the

*Table A-5 in the Labor Value charts (Appendix A) gives a value of 24.38 hours for 100 switches; .24 hour is therefore required for a single switch.

estimator writes the total labor time, 70 × .24, or 18. The entry now appears as shown in Table TO-9.

Conduit and wire are handled somewhat differently on the price sheet than other items. They are entered on the sheet in their normal unit of sale—100 feet or 1000 feet—in the quantity column. In Table TO-11, for example, 46,000 feet of ¾-inch rigid galvanized conduit is entered as 460. It then becomes a simple matter to multiply 460 by the cost per 100 feet given in the catalogues. Even when the required quantity is less than the usual unit of sale, this method is helpful. For instance, if 30 feet of 3½-inch rigid galvanized conduit is needed, it is entered as .3. The unit cost is $143.75, therefore the cost extension is .3 × $143.75, or $43.

For No. 14 THHN wire, 36,000 feet would be entered as 36 and multiplied by the unit cost of $22.80 per thousand feet for a total cost of $821. Similarly, 1700 feet of 250 MCM wire would be entered as 1.7 and multiplied by the unit cost of $1126, yielding $1914 for the cost extension. Cable is sold in units of 1000 feet and therefore should be handled like wire on the pricing sheet.

The estimator should be sure to enter the total quantities required of any given item, even though the item appears on more than one take-off sheet. For example, ¾-inch rigid galvanized conduit should be entered on the price sheet as a single item—that is, as the total of the subtotals for ¾-inch RGC on the several take-off sheets on which it appears. This practice should be observed for all items of the same description. To ensure that he has included all the items of the same description, the estimator should review the take-off sheets to see if all entries and subtotals for a particular item have been redded out.

The estimator arranges items on the price sheet in ascending size order. For instance, he starts the list of rigid galvanized conduit with the smallest size used on the job and works up to the largest. The ascending order matches the order in most labor value tables and therefore makes the estimator's work easier, faster, and more accurate. The estimator separates the groups of items (EMT, RGC, etc.) by spaces. (The space can be used for items that have been added or overlooked.)

He starts the list of lighting fixtures on the price sheet with the first type on the legend sheet—usually labelled "A". After he has completed the fixture price sheet (see Table TO-9), the estimator looks over his notes to see if every one of them is redded out. He then looks over the drawings once more to see if every note on the drawings has been redded out. If all is in order, he continues with the next step.

Labor Values

The best labor values for column F on the price sheet are those that the estimator's company has developed through experience over the years. New firms, of course, do not have such data, and small firms may not have the resources to gather the data. Fortunately, published data is available. Charts of certain labor values are contained in Appendix A of this book, for instance.

Some publications give "unit prices"—prices for specific operations, perhaps even for such details as tapping a nail into place. To use unit prices, the estimator must determine the operations that are required to install and wire a given item and find the total price for the operations.

Other publications print "labor values"—the typical times required to install and wire the various items of electrical hardware. These should be checked against past experience before they are used. The author prefers to use labor values rather than unit prices, since they can be used directly, without "translation." A unit price might not include some hidden item, or it might include an unnecessary part. Labor values exclusively are used in this book.

Whatever source the estimator uses for labor data, he should make it a point to understand the factors and conditions pertaining to the data, to ensure that the values are applicable to his job. He should not be afraid to temper published values with his own experience. (The components of the labor values provided in this book are discussed in Appendix A.)

The estimator finds the labor value for every item on the price sheet. If no labor value is required—if, for example, the item will be installed by another contractor—the estimator puts a dash in the labor space to show that the item has been considered. No estimating of the labor for lighting fixtures should be done without first looking at the manufacturer's catalogue. The catalogue contains detail and section drawings that will suggest quick and simple ways of mounting the fixtures. The catalogue drawings are particularly helpful when local authorities impose local requirements, such as mounting fixtures to the building frame.

Labor Rate

When the estimator has calculated the labor extensions for all the items, he adds them to arrive at the total labor hours for the entire job. He multiplies the total by the labor rate used by his company.

The labor rate should not be influenced by variables such as productivity, job conditions, weather, or unusual hazards. The variables will be taken into account on the summary sheet.

Material Costs

During the bid period, the estimator has presumably been talking to various suppliers about the items that will be needed for the job. Local electrical wholesalers are very helpful in this phase. A wholesaler can give the estimator the prices, from many manufacturers, for most of the items needed. It would be a major chore for the estimator to gather this information himself. However, a price quoted by a wholesaler is a "street price;" the estimator has no particular advantage because the wholesaler will quote the same price to all the bidders on a job.

Material Pricing Strategy

Wholesalers usually have to change their prices for conduit and wire as the prices for raw materials vary. However, the job will not be awarded for 30 to 60 days and the materials will not be needed for probably a year after that. If it seems likely that a price at construction time will be higher, the estimator can take advantage of the current price by ordering now. On the other hand, if it seems that a price will drop, the estimator may want to delay ordering. Alternatively the contractor can agree to "ceiling" and "floor" prices with the wholesaler.

When the estimator prices fixtures, he should get prices from more than one supplier. But he should not necessarily select the lowest-price supplier of each fixture. Instead, he should select the one supplier who can give him the lowest prices for *all* the fixtures. The estimator will thereby gain the advantage of a larger discount because he orders a large quantity from a single source instead of small quantities from many.

System Costs

For prices of the various systems—fire alarm, door security, nurses' call, and others — the estimator must go directly to the manufacturer. He should obtain several quotes, and ensure that the lowest-price system meets the specifications before he accepts it. The manufacturers will probably quote the same "street prices" to all, including the estimator's competitors.

For some systems, such as a lightning protection system, the electrical contractor may employ a subcontractor who specializes in that type of installation. Other systems, such as public address or intercom systems, may be purchased completely installed by others.

Regardless of who supplies the system and who installs it, the electrical contractor is responsible for it. Therefore all systems should be entered on the price sheet and the item cost and labor cost columns should contain either a dollar figure or a dash to show that the system has been included in the calculations.

"Eyeball" Checks

At this point, two simple and quick back-checks can be made to ensure that no major omissions have been made in the take-offs of (1) wire and (2) cable. The wire take-off is checked by comparing the total length of all wire to the total length of all conduit. For a three-wire circuit, there should be 3 feet of wire for every foot of conduit, plus an allowance for ups and downs, splices, and connections. Therefore the total wire should be at least 3 times the length of the conduit, but—unless there is a valid reason for it—not more than 4 times. A ratio of wire to conduit of, say, 3.4 would assure the estimator that his totals for conduit and wire are reasonable. For two-wire and four-wire circuits, the ratio should be slightly greater than 2 or 4, respectively.

The cable take-off can be "eyeball" checked by comparing the length of cable to the number of switches, receptacles, and lighting fixtures. The average footage per item is found by adding two or three representative receptacle runs, switch run, and fixture run and taking the average. The average footage per item multiplied by the total number of cable-connected switches, receptacles, and fixtures, should be approximately equal to the total amount of cable on the price sheet. (Alternatively, the total cable footage can be divided by the total number of cable-connected receptacles, switches, and fixtures. The estimator can then exercise his judgment to determine if the resulting average footage per item seems reasonable.)

If there is any doubt about the results of an eyeball check, more refined methods can be used. In addition, it is helpful to check the basic arithmetic in the pricing calculations.

Double Checking

An estimator should always have his calculations, no matter how simple, checked by another person. Another person can approach the take-off with a fresh eye and detect errors, both large and small, that the estimator might overlook through familiarity or repetition. Because many jobs have been lost—or won unprofitably—through errors in trivial details such as misplaced decimal points or commas, it bears repeating: another person must check all computations.

THE SUMMARY SHEET

The estimator uses the summary sheet to write the totals from the pricing sheets and to consider several factors that may affect the bid price.

The ideal summary sheet is one that a contractor has developed over the years—one that is oriented to his particular needs and includes all the items he requires. Less desirable, but nonetheless useful, are various printed summary sheet forms.

Whatever type is used, the summary sheet (also known as a recapitulation sheet) should be headed by the following information:

- Full name of project
- Address and location of job
- Estimator's name
- Checker's name
- Approver's name
- Estimate number
- Total number of sheets in the estimate
- Date
- Architect's name
- Engineer's name
- Owner's name
- Person to whom estimate is to be delivered
- Due date for estimate

For each pricing sheet, the estimator adds the entries in the "Extension of Item Cost" column. The result is the total dollar cost of the items on each sheet. Similarly, the estimator adds the entries in the "Extension of Labor Value" column for the total labor hours. The estimator takes these totals from the price sheets and transfers them to the summary sheet opposite their page number. A sample summary sheet appears in Table TO-12.

We will discuss in detail the many factors that the estimator should take into account on the summary sheet.

1. *Materials.* The total cost of materials and items that appears on the summary sheet. (Dollars.)*

2. *Sales tax.* The statutory tax imposed on the materials. (Dollars.)

3. *Labor (productive).* The hours spent in actual installation. This figure is the total labor that appears elsewhere on the summary

*The word dollars, hours, or percent in parentheses indicates the units in which an item should be carried.

sheet. (Hours.)

4. *Labor (nonproductive).* Labor charged by people who do not participate in the actual installation. This includes time charged to the job by such personnel as truck drivers, stock clerks, draftsmen, and engineers. It also includes time spent by a foreman in planning a layout or solving a mounting or circuitry problem. (Hours or dollars.)

5. *Supervision.* Time charged by foremen and others for supervising productive labor. (Hours, dollars, or percent.)

6. *Travel time.* Applicable to all who are paid for time spent in traveling to and from the site. May include productive labor people as well as others. (Hours, dollars, or percent.)

7. *Expenses.* Equipment and materials used exclusively for the job, but not installed—walkie-talkies, paper, pencils. Can also include maintenance costs such as expenditure for repairs to a field adding machine. (Dollars.)

8. *Temporary light and power.* The electric bill during construction, if the electrical contractor is responsible for it. May also include the bill during "beneficial use and occupancy" by the owner—when the owner moves in before the building is substantially (usually 98 percent) complete. If a lamp guarantee is part of the contract, cost and labor for replacing broken or burnt-out lamps must be included. (Dollars.)

9. *Temporary wiring for light and power.* The cost of materials and labor for temporary electrical wiring during construction and beneficial use and occupancy.

10. *Material handling—special.* Charges for special one-time handling that might be overlooked in the estimating procedure, not for normal movement of material. Example: disassembly of a switchboard and reassembly on another floor. (Dollars or percent.)

11. *Lost time.* Nonproductive time resulting from an unforeseeable event such as an accident or a jurisdictional dispute. (Dollars or percent.)

12. *Overtime.* Premium rate labor out of normal working hours. Overtime is carried as a separate item because it would distort the productive labor rate. (Hours.)

13. *Permit and inspection fees.* Charges imposed by the local government, sometimes based on the current rating of the entrance switch. (Dollars.)

14. *Security.* Includes charges for guards, locks, telephones, and lighting at security stations. (Dollars or percent.)

15. *Telephone.* For the electrical contractor's use in ordering materials and keeping in touch with the home office. (Dollars.)

16. *Rigging.* Equipment for lifting heavy items. Should also include charges for standby men if a union jurisdictional problem may arise.

17. *Direct job expense.* Charges for tools or equipment rented or purchased for this job only. If, however, the electrical contractor opts to purchase (instead of rent) a piece of equipment—say a 5-inch hydraulic bender—in order to do a job, he may pro-rate a portion of the cost to the job. (If the equipment has a 5-year life and it will be used on the job for a year, the estimator would charge 20 percent of the purchase price.) Direct job expense is a tricky item because it is easily confused with overhead. The difference is that overhead is for equipment that must be continually repaired or replaced. (Dollars or percent.)

18. *Testing.* Costs of supplying a standby man or factory-trained expert—or both—during testing or first operation of certain equipment. Cost of oil for testing the emergency generator, plus cost of oil to refill the tank. Cost of electricity for running tests. (A 1200-horsepower air conditioning unit operating for five days at eight hours a day can run up a sizable electric bill.) The specifications usually spell out the responsibilities; it is possible that another contractor may be required to assume all or part of the testing expense. (Hours or dollars.)

19. *Productivity.* A factor that enables the estimator to re-assess his original labor rate. By now he is more familiar with the job and may discern conditions that increase or decrease the difficulty of the work, and can raise or lower the labor rate accordingly. An indoctrination or learning period may be necessary, for example. (Percent.)

20. *Complexity.* Like productivity, this item allows the estimator to make adjustments. In this case, he can adjust his total up or down, depending on the complexity of the job viewed as a whole. It may be necessary to coordinate efforts at different levels of the building, for example, during installation of a system. (Percent.)

21. *Coordination.* Charges for coordinating work with other contractors. Includes cost of electrical layout drawings supplied to others.

22. *Size of job.* Extra costs that might accrue from movement of people and materials over large distances within the job site—if the job is a high-rise, for instance, or if the work shack is far from the construction. Can also be used to carry costs saved if the job is unusually compact. (Percent.)

23. *Weather.* Added cost if bad weather is expected during the construction period or if the work will be done in a hot climate. The Weather Bureau can advise the estimator here. (Percent.)

24. *As-built drawings.* Costs of drawings furnished to the owner showing how the wiring was actually installed.

25. *Escalation.* A charge intended to compensate the bidder for increases in the price of materials during the time the award is pending as well as during construction. Calculated as a percent of the original

bid per month for a period extending from the date of submission of the bid to the date when construction is half finished (1½ percent escalation for 24 months means a total escalation of 36 percent). Also includes reductions in cost, or de-escalation.

26. *Bond.* Charge for posting a performance bond. (Percent.)

27. *Insurance.* Workmen's compensation, unemployment insurance, and similar taxes are usually combined and carried as percentage of the hourly labor rate. Similar insurance for nonproductive employees is usually included in the percentage, but can be carried separately, if desired, under overhead. Fire, theft, accident, and other insurance for the job is carried as a percentage of direct job expense.

28. *Overhead.* Business expenses that are not chargeable to a specific job—the normal costs (or burden) of day-to-day operation of the business. Includes rent, light, heat, telephone, taxes, office salaries, advertising, and whatever other costs the contractor or his accountant or lawyer feels are appropriate. (Percent.)

29. *Amount of estimate.* The sum of the preceding items, after they have been converted into dollars if given as hours or percent. To convert hours, the estimator multiplies by the hourly labor rate. To convert percent, he takes the percentage of the total price before profit is added. (Dollars.)

30. *Profit.* A percentage of the amount of the estimate to be added to the amount. Many estimators tend to use profit as a "fudge factor," adding a few percentage points to compensate for various corrections (such as items 10, 19, 20, and 21 on the check list) which they have neglected. It is a far better practice to include corrections in the estimate and thereby obtain a truer picture of the elements involved in the overall cost—and maintain a reasonable, proper, and realistic profit margin. (Percent.)

31. *Price total.* The amount of the estimate plus, after conversion to dollars, the profit. (Dollars.)

32. *Real purchasing price.* Here, the estimator determines what discounts are available to him and what portion of them he wishes to pass on to the customer. (Dollars.)

33. *Real productivity.* A final check of productivity, Item 19.

34. *Price submitted.* The price to be quoted to the customer.

Submitting the bid

There are many details that must be attended to after the price has been decided and before the bid is submitted. First, the precise time and place for submitting the bid should be determined. If the bid will be mailed, it should be sent—well in advance of the due date—by registered mail so that a signed receipt is obtained. If a deposit check is required, it should be of the proper type (money order, cashier's check, or whatever is required), in the correct amount, and payable to the proper name. Forms or cover letters should be signed by the proper person.

If the bid is to be delivered personally, two people should accompany it. In case any last minute problems crop up, one person can stay with the bid while the other can take whatever action is necessary. Trivial as it may seem, the couriers should have ample change for parking and telephone calls—it could be crucial to turning in the bid on time and in the correct form if the bid opening is to be public.

If, after the bids have been opened, another company is found to be the low bidder in public bidding the courier should examine the low bid for possible errors and omissions. If any are found, the awarding authority should be informed, since the apparent low bidder may be disqualified.

The winning bidder, if his estimate has been prepared according to the procedure described in this book, can immediately use his estimate as material and quantity ordering sheets.

APPENDIX A
LABOR VALUES

A labor value is the measure of the time involved in performing a given task, such as installing a certain type of lighting fixture or installing 100 feet of rigid galvanized conduit. There is some question as to when such a task begins. Does it begin when the electrician picks up the item at his work location and starts to install it? Or does it begin when the required materials are taken off the truck at the construction site? Or when the electrical contractor orders the materials from a supplier?

As a result of this ambiguity, there are two definitions of labor value. One definition, suitable for large contractors, includes—in addition to the work of actually installing the item—the work of bringing the item from the truck tailgate to the point of installation. For example, suppose a fixture is to be installed on the twentieth floor of a building. A journeyman unloads the fixture from the truck and brings it to a central staging area for electrical goods. Another journeyman selects it and brings it to the twentieth floor. Once there, the container and wrapping must be removed. Finally, the fixture is assembled and mounted.

The other definition—suitable for smaller contractors—includes all of the above plus the time for ordering and picking up items from suppliers.

The reason for the two definitions is that large contractors have a staff devoted to purchasing whose time is charged to overhead. In addition, large contractors order in large enough quantities and sufficiently ahead of time that suppliers can include delivery costs in their prices. Often, materials will be shipped directly from the manufacturer.

Small contractors, on the other hand, do not enjoy these advantages. They do not have a separate purchasing staff, and frequently must pick up materials themselves. The small contractor, therefore, should use a higher labor value than the large contractor.

Almost all labor charts and tables indicate the conditions of delivery on which they are based. The labor values on the following pages are based on materials delivered to the job site at a normal truck delivery point. They *do not include* work expended in bringing the materials to the site. They are therefore suitable for use by large contractors. They are also suitable for the smaller contractor, provided delivery costs are added to the estimate. Delivery costs can be calculated precisely or—more conveniently—they can be estimated by adding 25 percent of the productive labor cost to the bid.

The author has developed the values from his own experience. He has compared them with the values in the many labor tables already in print. Whenever possible, he has combined items of a similar nature in a single table. For example, the labor values for ½-inch conduit—EMT, PVC, rigid aluminum, and rigid galvanized—are arranged on one sheet for easy and immediate comparison.

The modular motor table includes all the equipment and devices for a motor installation. The table furnishes the total labor value for given horsepower and voltage.

The following tables, like any other labor tables, should be regarded as guides. The estimator is free to adjust the values up or down as his experience dictates.

TABLE A-1 CONDUIT LABOR VALUES

SIZE (INCHES)	RIGID GALVANIZED HW	ALUMINUM RIGID HW	ELECTRICAL METALLIC TUBING	POLYVINYL CHLORIDE
½	4.75	4.75	4.00	4.50
¾	5.65	5.90	4.75	5.50
1	7.45	7.25	6.40	6.75
1 ¼	8.70	7.45	7.10	7.00
1 ½	10.30	8.85	8.00	7.25
2	13.50	11.00	9.70	10.25
2 ½	18.45	13.90	11.85	12.75
3	22.55	17.00	14.00	15.00
3 ½	27.25	20.40	17.00	18.00
4	35.70	25.15	18.00	24.50
5	50.10	36.70		33.00
6	65.00	49.00		44.00

Hours for 100 ft. of installation
Includes fastenings, mounting, and terminating.

TABLE A-2 WIRE PULLING LABOR VALUES

SIZE OF WIRE (GAUGE)	COPPER		ALUMINUM	
	SOLID	STRANDED	SOLID	STRANDED
16	5.00	6.00		
14	6.25	6.90		
12	7.65	8.00	6.25	7.25
10	9.00	10.40	9.00	10.00
8		11.95		11.00
6		13.05		12.00
4		16.40		14.00
3		18.00		
2		18.70		15.25
1		19.55		16.25
1/0		24.60		20.00
2/0		27.65		21.25
3/0		30.00		24.25
4/0		36.70		27.75
250MCM		37.70		28.75
300MCM		42.10		30.75
350MCM		43.50		32.75
400MCM		44.00		34.75
500MCM		50.00		36.00
600MCM		57.00		41.50
750MCM		64.00		46.00
1000MCM		78.00		55.50

Hours per 1000 ft. of installation.
All insulation is valued using Thermoplastic Type insulation. #16-10 are computed as "branch circuits," less length and more 90's. #8-1000MCM evaluated as feeder and figured on an average of 150 ft. length per run. All wires are left hanging in their panels or boxes.

TABLE A-3 CONDUIT AND WIRE LABOR VALUES (LABOR VALUES COMBINED INTO ONE LABOR VALUE)

CONDUIT SIZE/ GAUGE OF WIRE	ELECTRIC METALLIC TUBING		RIGID GALVANIZED CONDUIT	
	3 WIRE	4 WIRE	3 WIRE	4 WIRE
½″/14	6.28	7.03	7.02	7.79
½″/12	6.64	7.52	7.39	8.27
½″/10	7.43	8.58	8.18	9.33
½″/8	7.94		8.69	
¾″/8		10.01		10.91
1″/6	10.71	12.14	11.75	13.20
1″/4	11.81		12.86	
1¼″/4		14.32		15.92
1¼″/3	13.04	15.02	14.64	16.62
1¼″/2	13.27	15.33	14.87	16.93
1¼″/1	13.55	16.60	15.15	18.90
1½″/0	16.12		18.42	
1½″/00	17.12		19.42	
2″/0		20.52		24.32
2″/00		21.87		25.67
2″/000	19.60	22.90	23.40	26.70
2″/0000	21.81		25.61	
2½″/0000		28.00		34.60
2½″/250MCM	23.99	28.14	30.89	35.04
2½″/300MCM	25.44		32.34	
3″/300MCM		32.52		41.07
3″/350MCM		33.14		41.69
3″/400MCM	28.52	33.36	37.07	41.91
3″/500MCM	30.50		39.05	
3″/600MCM	32.81		41.36	
3½″/500MCM		39.00		49.25
3½″/600MCM		42.08		52.33
3½″/750MCM	38.12		48.37	
4″/750MCM		46.16		63.86

Units are based on 100 ft. of conduit, the number of wires in the conduit times 100, and 10% for tying in, etc.

TABLE A-4 ARMORED AND NONMETALLIC CABLE LABOR VALUES

GAUGE OF CONDUCTORS	ARMORED CABLE			NONMETALLIC	
	2 WIRE	3 WIRE	4 WIRE	2 WIRE	3 WIRE
14	2.49	2.83	3.50	2.36	2.73
12	2.77	3.17	4.00	2.73	3.12
10	3.78	4.13	5.00	3.62	4.38
8	4.94	5.19	7.50	4.60	5.57
6	5.30	5.50	8.80	6.13	
4	6.40	6.75	10.60		
2	6.80	7.00	11.20		
1	7.90	8.25	13.00		
1/0	8.30	8.50	13.80		
2/0	9.40	9.75	15.60		
3/0	10.00	10.25	15.90		
4/0	11.50	11.75	18.70		
250MCM	12.00	12.50	20.00		
350MCM	13.00	13.75	21.80		
500MCM	14.00	15.50	23.80		
750MCM	16.50	18.00	27.50		

Hours per 100 ft. of cable.

TABLE A-5 DEVICE LABOR VALUES

	AMPERE RATING 15 (#14 wire)	AMPERE RATING 20 (#12 wire)	AMPERE RATING 30 (#10 wire)	AMPERE RATING 50 (# 6 wire)
Single pole switches S1	19.00	24.38	31.67	
Double pole switches S2	39.40	51.00	61.67	
Three way switches S3	30.40	35.67	46.66	
Four way switches S4	39.40	51.00	61.67	
Duplex receptacles	21.30	25.38		
Duplex receptacles, 2 cir.	31.50	34.00		
Single receptacles, 115/250 volt 3 pole	30.17	39.88	50.70	66.25
Single receptacles, 115/250 3 pole 60Amp				*92.50
Clock receptacle	32.50			
Fan receptacle	51.50			
Plates, per gang	8.66			
Weatherproof plates	12.50			

*60 Amp
Units figured at 100 pieces per hour.

TABLE A-6 PANEL LABOR VALUES

PANEL SIZE (AMPERES)	TWO POLES	THREE POLES	FOUR POLES
30	2.45	2.57	2.85
100	3.58	4.06	5.02
225	4.70	6.82	7.91
400	9.18	10.98	13.13
600	12.28	15.10	17.32
800	18.63	21.29	24.60
1000	20.82	23.65	25.80
1200	23.00	27.00	29.00

Labor figures include unpacking, removing the trim, removing the interior, storing both the trim and interior and mounting the back box in place. Fastenings are included.

TABLE A-7 CONNECTIONS TO LUGS IN PANELS AND SWITCHES

SIZE OF BREAKER	SIZE OF CONDUCTOR WITH T-TW INSULATION	TWO CONNECTIONS	THREE CONNECTIONS	FOUR CONNECTIONS
15-40	14-8	.65	.96	1.26
55-95	6-2	.85	1.46	1.89
110	1	1.43	2.07	2.60
125	1/0	1.73	2.39	3.03
145	2/0	1.93	2.62	3.32
165	3/0	2.13	2.86	3.65
195	4/0	2.32	3.09	3.94
215	250MCM	2.58	3.43	4.42
240	300MCM	2.82	3.85	4.95
260	350MCM	3.05	4.24	5.47
280	400MCM	3.33	4.62	6.01
320	500MCM	3.64	5.01	6.61
355	600MCM	4.04	5.51	7.28
400	750MCM	4.65	6.06	8.03
455	1000MCM	5.62	7.25	9.03

Hours per unit.
Labor includes replacing interior, trim, and directory, plus lacing and stripping conductor and connecting.

TABLE A-8 MOTOR AND DISCONNECT SWITCHES, EXTERNALLY OPERABLE (250 VOLTS)

HP	SWITCH SIZE, AMPERES, AT 250 VOLTS	TWO CONNECTIONS	THREE CONNECTIONS	FOUR CONNECTIONS
Fractional	Toggle motor rated	1.99	2.20	2.31
1-3	30	2.17	2.33	2.45
5	30	2.24	2.47	2.60
7.5	60	2.50	2.96	3.34
10	60	2.65	3.14	3.54
15	60	2.81	3.33	3.75
20	100	3.16	3.76	4.50
25	100	3.35	3.99	4.77
30	100	3.55	4.23	5.06
40	200	5.43	6.36	7.31
50	200	5.98	7.00	8.04
60	200	6.58	7.70	8.84
75	400	8.68	10.53	12.47
100	400	9.55	11.58	13.72

Hours per unit.
Labor includes securing, mounting, fastening, connecting, and identifying. Essentially "ready for use."

TABLE A-9 MOTOR AND DISCONNECT SWITCHES, EXTERNALLY OPERABLE (600 VOLTS)

HP	SWITCH SIZE, AMPERES, AT 600 VOLTS	TWO CONNECTIONS	THREE CONNECTIONS	FOUR CONNECTIONS
Fractional	Toggle motor rated	2.36	2.72	3.21
1-3	30	2.50	2.88	3.40
5	30	2.65	3.05	3.61
7.5	30	2.82	3.23	3.83
10	30	3.10	3.65	4.25
15	30	3.29	3.87	4.50
20	60	3.49	4.10	4.78
25	60	3.70	4.35	5.07
30	60	3.92	4.78	5.37
40	100	4.31	5.26	6.11
50	100	4.74	5.79	6.72
60	100	5.22	6.37	7.39
75	200	6.65	8.20	9.22
100	200	7.13	9.02	10.14

Hours per unit.
Labor includes securing, mounting, fastening, connecting, and identifying.
Essentially "ready for use."

TABLE A-10 MAGNETIC ACROSS-THE-LINE STARTERS

CONT. CURRENT RATING	STARTER SIZE	*TW WIRE SIZE, CU	MAX. HP	TWO CONN. 230 VOLT	THREE CONN. 230 VOLT	THREE CONN. 480 VOLT	DEV
9	00	14	Fract-1	2.52			
9	00	14	1½-2		2.82	2.82	
18	0	14	1-2	2.52			
18	0	14	3		2.82		
18	0	14	5			2.82	
27	1	12	3	2.60			
27	1	10	7½		3.12		
27	1	12	10			2.94	
45	2	6	7½	3.27			
45	2	6	15		3.52		
45	2	6	25			3.52	
90	3	1	30		4.49		
90	3	6	30			3.52	
90	3	4	40			4.07	
135	4	2/0	40		4.97		
135	4	3/0	50		5.27		
90	3	2	50			4.37	
270	5	4/0	60		7.22		
135	4	1	60			4.49	
270	5	300MCM	75		7.82		
135	4	1/0	75			4.67	
270	5	500MCM	100		8.72		
135	4	3/0	100			5.27	
540	6	750MCM	125		16.86		
270	5	4/0	125			7.22	
540	6	1000MCM	150		17.16		
270	5	300MCM	150			7.82	
540	6	2-500MCM	200		21.36		
270	5	500MCM	200			8.72	

Mounted in starter, not wired:

Stop or start station 3

Stop and start station 4.

Stop and start station with pilot light 7.

Remote stop or start station 1.3

Remote stop and start station 1.4

Remote stop and start station with pilot light 2.0

Hours per unit.

Labor includes stripping, connecting, drilling, mounting, fastening NEMA 1 enclosures. Line and load connections are included. Also 2 heaters.

*Wire size computed at 125% of full load of motor.

TABLE A-11 CONNECTIONS AT MOTORS

MAX. HP MOTOR	TW INSUL, WIRE SIZE, CU	TWO POLE 230 VOLT	THREE POLE 230 VOLT	THREE POLE 460 VOLT
Fract-1	14	.43		
1½-2	14		.65	.65
1-2	14	.43		
3	14		.65	
5	14			.65
3	12	.54		
7½	10		.97	
10	12			.83
7½	6	.97		
15	6		1.46	
25	6			1.46
30	1		2.20	
30	6			1.46
40	4			1.62
40	2/0		3.73	
50	3/0		4.21	
50	2			1.78
60	4/0		4.86	
60	1			2.20
75	300MCM		6.38	
75	1/0			3.24
100	500MCM		8.26	
100	3/0			4.21
125	750MCM		12.96	
125	4/0			4.86
150	1000MCM		15.00	
150	300MCM			6.38
200	2-500MCM		16.72	
200	500MCM			8.26

Hours per unit.
Labor includes stripping, connecting, taping, either split conn or lugs in motor terminal box. Motor is mounted in place by others.

TABLE A-12 DRY TYPE TRANSFORMERS: 480-240 PRIMARY, 240-120 SECONDARY

		PRIMARY				SECONDARY				1φ	1φ	3φ	
		480		240		240		120		FLOOR	WALL	FLOOR	W.
KVA	φ	W	A	W	A	W	A	W	A	MOUNTED	MOUNTED	MOUNTED	MOU
.50	1	14	2			14	4			2.93	2.93		
.50	1			14	3			14	6	2.93	2.93		
.75	1	14	2			14	4			3.18	3.18		
.75	1			14	4			14	7	3.18	3.18		
1.00	1	14	3			14	6			3.43	3.81		
1.00	1			14	5			14	9	3.43	3.81		
1.50	1	14	4			14	7			3.68	4.15		
1.50	1			14	7			14	13	3.68	4.15		
2.00	1	14	5			14	9			3.93	4.62		
2.00	1			14	9			12	17	3.98	4.67		
3.00	1	14	7			14	13			4.21	5.09		
3.00	1			14	13			10	25	4.48	5.35		
3.00	3	14	4			14	8					4.93	[illegible]
5.00	1	14	11			10	21			5.27	6.60		
5.00	1			10	21			6	42	5.54	6.88		
6.00	3	14	8			14	15					6.43	[illegible]
7.50	1	12	16			8	32			6.63	7.69		
7.50	1			8	32			4	63	7.42	8.48		
9.00	3	14	11			10	22			8.50	9.87		
10.00	1	10	21			6	42			8.87	10.58		
10.00	1			6	42			2	84	9.42	11.13		
15.00	1	8	32			4	63			10.96	14.32		
15.00	1			4	63			1/0	125	12.36	15.72		
15.00	3	12	19			8	37					10.40	12
25.00	1	6	53			1	105			14.54	19.96		
25.00	1			1	105			250	209	17.42	20.57		
30.00	3	8	37			3	74					12.23	14
37.50	1	3	79			3/0	158			19.91	25.65		
37.50	1			3/0	158			[2]3/0	316	24.19	29.93		
45.00	3	6	55			1	110					19.07	21
50.00	1	1	105			250	210			24.02	30.13		
50.00	1			250	210			[2]250	420	30.25	36.36		
75.00	1	3/0	157			500	314			30.02	38.82		
75.00	1			500	314			[2]500	628	38.23	47.03		
75.00	3	2	91			4/0	182					26.69	29
100.00	1	250	209			[2]250	418			38.05	43.75		
100.00	1			[2]250	418			[3]400	836	53.10	58.80		
112.50	3	2/0	136			400	272					37.43	43
150.00	3	4/0	181			[2]4/0	362					43.22	49
167.00	1	600	348			3/300	698			70.44	76.74		
167.00	1			3/300*	698			5/400	1396	88.28	94.58		
225.00	3	400	271			2/400	542					65.85	81
300.00	3	700	362			3/350	724					83.22	102

*3/300=sets of conductors/size of conductor

Hours per unit.
KVA=Kilovolt-Amperes
W=Conductor size
A=Ampere
φ=Phase

Labor includes mounting, fastening, wire connection (split bolt or lug), and taping. Essentially "ready-to-use."

TABLE A-13 LIGHTING FIXTURES LABOR VALUES

LENGTH OF FIXTURE	NUMBER OF LAMPS	DESCRIPTION	MOUNTED CEILING	MOUNTED RECESSED	MOUNTED PENDANT
4	1	Strip,bare	1.24	1.75	1.55
4	2	Strip, bare	1.39	1.89	1.69
4	4	Strip, bare	2.06	2.16	2.36
4	2	Strip, w/lens	1.74	2.25	2.05
4	4	Strip, w/lens	2.56	2.66	2.86
4	2	Troffer	2.10	2.60	
4	4	Troffer, w/lens	3.05	3.16	
	Hi Hat		1.20	1.49	1.70

Hours per fixture.
Labor includes mounting, fastening, connecting, and lamping. Essentially "ready to use."

TABLE A-14 UNIT OR MODULAR MOTOR LABOR VALUES

HP	POLES	VOLTS	WIRE SIZE TW, CU	STARTER SIZE	STARTER LABOR	DISC.SW. SIZE	DISC.SW. LABOR	MOTOR CONN. LABOR	TOTAL LABOR
Fract	2	110-230	14		1.99	Toggle		.43	2.42
Fract	3	230	14		2.20	Toggle		.65	2.85
Fract	2	460	14		2.36	Toggle		.43	2.79
Fract	3	460	14		2.72	Toggle		.65	3.37
Fract-1	2	230	14	00	2.52	30	2.17	.43	5.12
1½-2	3	230	14	00	2.82	30	2.17	.65	5.64
1½-2	3	460	14	00	2.82	30	2.45	.65	5.92
1-2	2	230	14	0	2.52	30	2.17	.43	5.12
3	3	230	14	0	2.82	30	2.33	.65	5.80
3	2	230	12	1	2.60	30	2.17	.43	5.20
5	3	460	14	0	2.82	30	2.47	.65	5.94
7½	2	230	6	2	3.27	60	2.50	.97	6.74
7½	3	230	10	1	3.12	60	2.96	.97	7.05
10	3	460	12	1	2.94	30	3.65	.83	7.42
15	3	230	6	2	3.52	60	3.33	1.46	8.31
25	3	460	6	2	3.52	60	4.35	1.46	9.33
30	3	230	1	3	4.49	100	4.23	2.20	10.92
30	3	460	6	3	3.52	60	4.78	1.46	9.76
40	3	460	4	3	4.07	100	5.26	1.62	10.95
40	3	230	2/0	4	4.97	200	6.36	3.73	15.06
50	3	230	3/0	4	5.27	200	7.00	4.21	16.48
50	3	460	2	3	4.37	100	5.79	1.78	11.94
60	3	230	4/0	5	7.22	200	7.70	4.86	19.78
60	3	460	1	4	4.49	100	6.37	2.20	13.06
75	3	230	300MCM	5	7.82	400	10.53	6.38	24.73
75	3	460	1/0	4	4.67	200	8.20	3.24	16.11
100	3	230	500MCM	5	8.72	400	11.58	8.26	28.56
100	3	460	3/	4	5.27	200	9.02	4.21	18.50

Devices mounted in starter, not wired	
Stop or start station	.30
Stop and start station	.45
Stop and start station w/pilot light	.75
Remote stop or start station in NEMA 1 enclosure	1.30
Remote stop and start station in NEMA 1 enclosure	1.45
Remote stop and start station in NEMA 1 enclosure w/pilot light	2.00

Labor values in hours.
Essentially "ready to use."

APPENDIX B EVALUATING CHANGE ORDERS

A change order is a change in the original contract document. The change may be necessitated by an error or omission—the structural load calculations may have been wrong or the electrical ampacity may be inadequate. Or there may be esthetic reasons for the change—perhaps the owner is dissatisfied with some aspect of the building's appearance.

Change orders are taken off by the same procedure used in the original estimate. The price of the change order is then compared with that of the relevant portion of the original estimate. Depending on whether the change costs more or less than the original, there will be a charge or credit to the customer.

It is important to prevent duplication of material and effort that might result from the change. This is done by marking on the drawings those items that are omitted by the change. The take-off for the change order is then checked to make sure the items have not been included.

If a change order involves a complete motor branch circuit or complete drawing, the estimator can review the original take-off sheets to find the quantities and labor values involved in his original estimate. However, if the change order is only partial, it is necessary to repeat the take-off of the affected conduit, wire, equipment, etc., shown in the original drawings. Next, the estimator does a take-off of the conduit, wire, equipment, etc. as shown on the change order.

He determines the prices and labor values for both take-offs. He adds an appropriate overhead and profit factor to the change order. He does not deduct profit or overhead from the original estimate.

The estimator makes up a summary sheet with the following four column headings: "Original," "Change Order," "Add," and "Deduct." On the summary sheet, he shows the original amount, the change order amount, and—depending on whether the change order amount is greater than or less than the original amount—the amount to be added or deducted from the original total investment. If there are several change orders, simple addition of the "Add" and "Deduct" columns will give the estimator the total amounts to be added to and deducted from the total.

APPENDIX C
DRAWINGS FOR A SERVICE AND SUPPLY BUILDING

Symbol	Description
A 3b	STANDARD DESIGNATION FOR ALL LIGHTING FIXTURES. A= FIXTURE TYPE, NUMERAL = CIRCUIT NUMBER, b = SWITCH CONTROL
	CEILING OUTLET W/FLUORESCENT FIXTURE "P" INDICATES PENDANT MOUNTED
	WALL OUTLET W/FLUORESCENT FIXTURE
	CEILING OUTLET W/INCANDESCENT FIXTURE
	WALL OUTLET W/INCANDESCENT FIXTURE
	FIXTURES WIRED ON EMERGENCY CIRCUIT
EX	EXIT SIGN. WITH ARROW AS INDICATED
	OUTLET BOX WITH EXTENSION RING
	FIXTURE PENDANT MOUNTED WITH UNISTRUT.
	FIXTURE MOUNTED ON RIB WITH SLIDE GRIP HANGERS
S	SINGLE POLE, SINGLE THROW SWITCH. MH. = 48" ℄
S2	DOUBLE POLE, SINGLE THROW SWITCH. M.H. = 48" ℄
S3	SINGLE POLE, DOUBLE THROW SWITCH. M.H. = 48" ℄
S4	DOUBLE POLE, DOUBLE THROW SWITCH. M.H. = 48" ℄
S.P.	SINGLE POLE SWITCH W/PILOT LIGHT, M.H. = 48" ℄
	SEALING GLAND
T	TRANSFORMER. 1 φ SPECIALTY TYPE 1500 WATT 277/120 V. WITH INTEGRAL FUSING.
	DUPLEX RECEPTACLE. M.H. 18" TO ℄
X	DUPLEX RECEPTACLE, "X" INDICATES ABOVE COUNTER M.H.= 42" OR ABOVE COUNTER
XP	EXPLOSION PROOF RECEPTACLE. M.H. = 36" TO ℄.
E	EMERGENCY CONDUIT AND WIRING
	PLUG IN BUS DUCT.
	PLUGMOLD STRIP.
D	INTRUSION ALARM; DOOR SWITCH
WM	WATCHMANS TOUR STATION
K	INTRUSION ALARM KEY RESET SWITCH WITH ADJACENT ALARM BELL
WM	WATCHMANS TOUR STATION W/ SECURITY PHONE IN NEMA 1 ENCLOSURE
P	20A 2GANG RECEPTACLE MOUNTED ON PLUG MOLD STRIP
P	40A-3P RECEPTACLE MOUNTED ON PLUG MOLD STRIP
	PLUG MOLD 2000 OR EQUAL WITH OUTLETS 18" O.C.
	CONDUIT STUB-UP
F	FIRE ALARM MANUAL STATION. M.H. = 48" TO ℄
	FIRE ALARM HORN WITH INTEGRAL WARNING LIGHT
	FIRE ALARM THERMODETECTOR, FIXED TEMPERATURE.
	20A-1P DISCONNECT SWITCH AND JUNCTION BOX
	FIRE ALARM THERMODETECTOR - RATE OF RISE

ABBREVIATIONS

M.H. MOUNTING HEIGHT
A.F.F. ABOVE FINISH FLOOR
U.H. UNIT HEATER
E.F. EXHAUST FAN
T.C.P. TEMPERATURE CONTROL PANEL
B.F.D. BOSTON FIRE DEPTARTMENT
E.C. ELECTRICAL SUB CONTRACTOR
U.V. UNIT VENTILATOR
C.B. CIRCUIT BREAKER
H.O.A. HAND-OFF-AUTOMATIC
E.W.C. ELECTRIC WATER COOLER
W.P. WEATHER PROOF
R.E.F. ROOF EXHAUST FAN
H&V HEATING AND VENTILATING

FIRE ALARM MASTER BOX. M.H. = PER BFD
SMOKE DETECTOR
MAGNETIC DOOR HOLDER
FIRE ALARM ANNUNCIATOR
ELECTRIC REHEAT COIL
PNEUMATIC ELECTRIC SWITCH - SUPPLIED BY OTHERS
ELECTRIC PNEUMATIC SWITCH - SUPPLIED BY OTHERS
TELEPHONE OUTLET. M.H. = 18" TO ℄
TELEPHONE AND COMMUNICATIONS CABINET
MOTOR - NUMERAL INDICATES HORSEPOWER
MOTOR STARTER, SUPPLIED AND INSTALLED BY E.C.
MOTOR DISCONNECT SWITCH 20A-3P UNFUSED UNLESS OTHERWISE NOTED
BUS DUCT JUNCTION BOX
JUNCTION BOX
CONDUIT TURNING UP
CONDUIT TURNING DOWN
CONDUIT & WIRE - RUN CONCEALED. № OF SLASHES INDICATE № OF WIRES, WHEN MORE THAN 2. WIRE SIZE TO BE #12 MIN. UNLESS NOTED OTHERWISE.
1,3,5 HOMERUN TO PANELBOARD. "LS" INDICATES PANEL DESIGNATION, 1,3,5, INDICATES CIRCUIT BREAKER No. (ALL BREAKERS 20A-1P UNLESS OTHER WISE NOTED) MINIMUM CONDUIT AND WIRE. 4 #12 + G - 3/4" CONDUIT UNLESS OTHERWISE NOTED
HEATING PANEL 277/430
LIGHTING PANEL 277/480
POWER PANEL 120/208
CLOCK CIRCUIT, CONDUIT AND WIRE
SNOW MELTING MAT.
CLOCK OUTLET WITH FLUSH CLOCK
ELECTRIC CABINET UNIT HEATER NUMERALS INDICATE KW
LIGHTNING ROD
LIGHTNING PROTECTION SYSTEM CONDUCTOR
TELEPHONE TERMINAL CABINET SUPPLIED AND INSTALLED BY THE ELECTRICAL SUB-CONTRACTOR.
FREEZE PROTECTION THERMOSTAT SUPPLIED AND INSTALLED BY HV-AC CONTRACTOR WIRED BY THE ELECTRICAL SUB CONTRACTOR.
DUCT MOUNTED SMOKE DETECTOR SUPPLIED AND INSTALLED BY HV-AC CONTRACTOR WIRED BY THE ELECTRICAL SUB CONTRACTOR
MOTORIZED GATE VALVE SUPPLIED AND INSTALLED BY PLUMBING SUB CONTRACTOR WIRED BY THE ELECTRICAL SUB-CONTRACTOR
METHANE PROBE SEE DETAIL DWG. E3-2
METHANE SLAB PROBE SEE DETAIL DWG. E3-2
WIRING IN PLUGMOLD
TELEPHONE CONDUIT, 3/4" RIGID STEEL WITH #12 FISHWIRE UNLESS OTHERWISE NOTED

DRAWING E1-1 SYMBOL LIST (LEGEND)

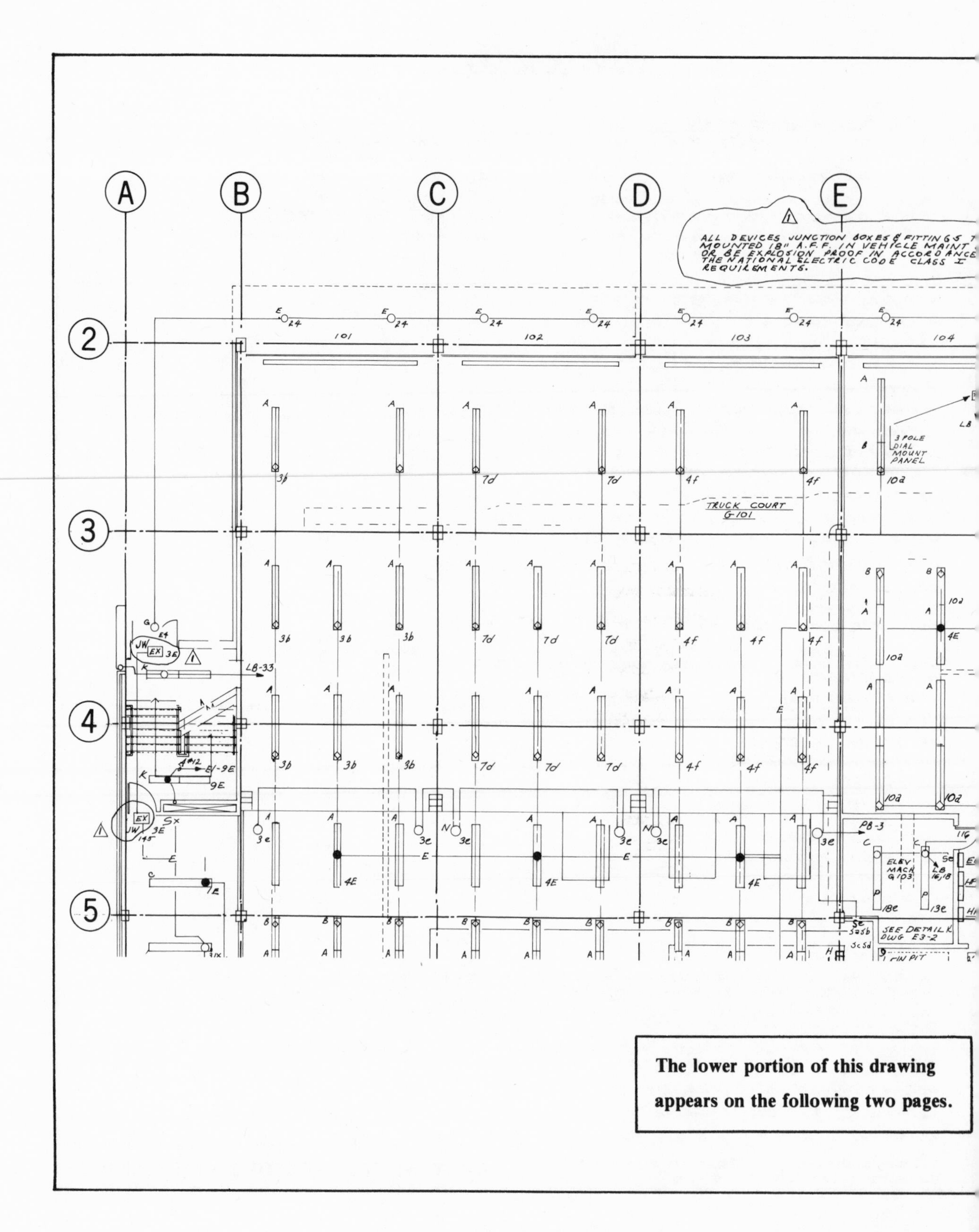

The lower portion of this drawing appears on the following two pages.

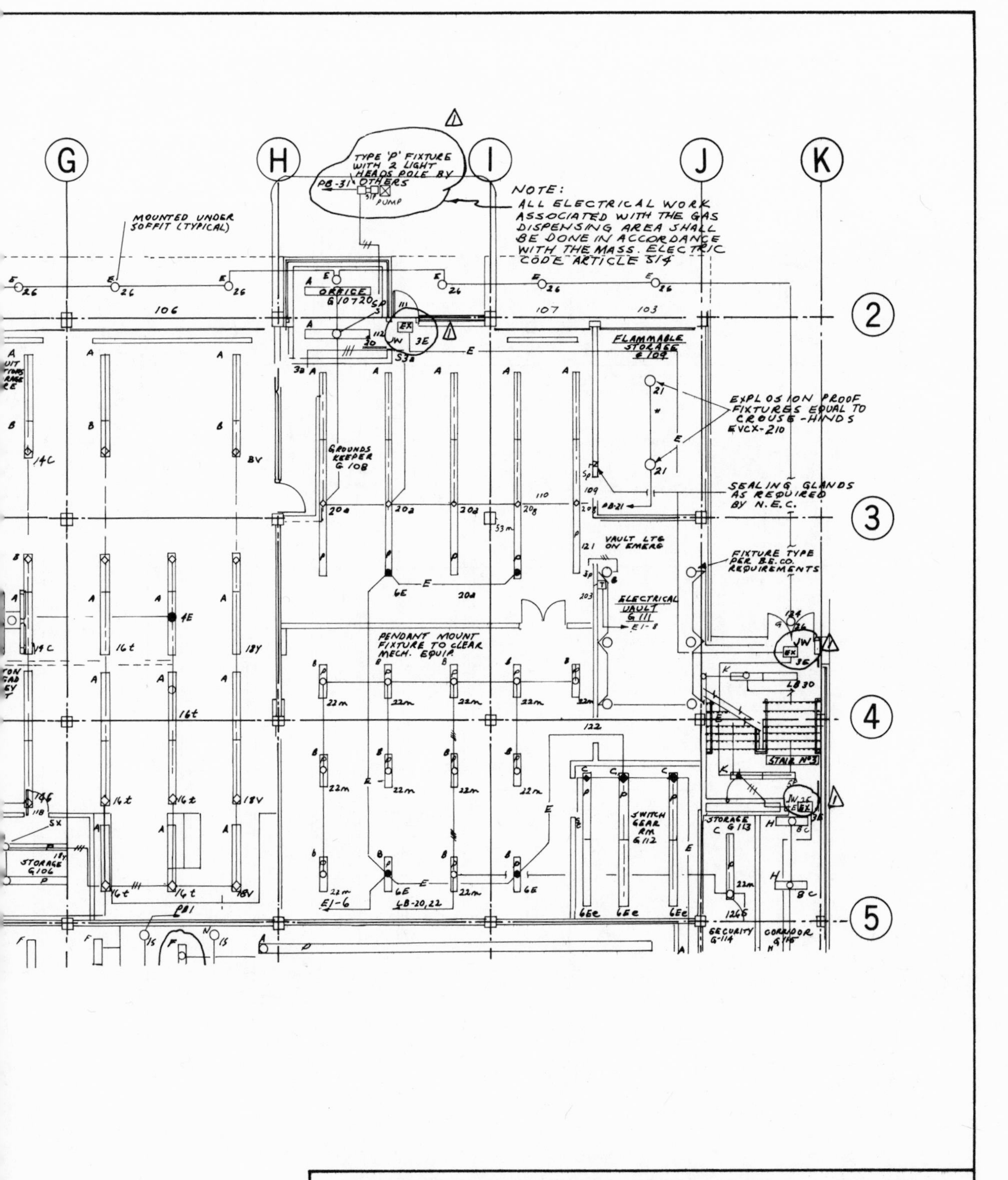

DRAWING E2-1 LIGHTING, LEVEL 24 (Continued Next Page)

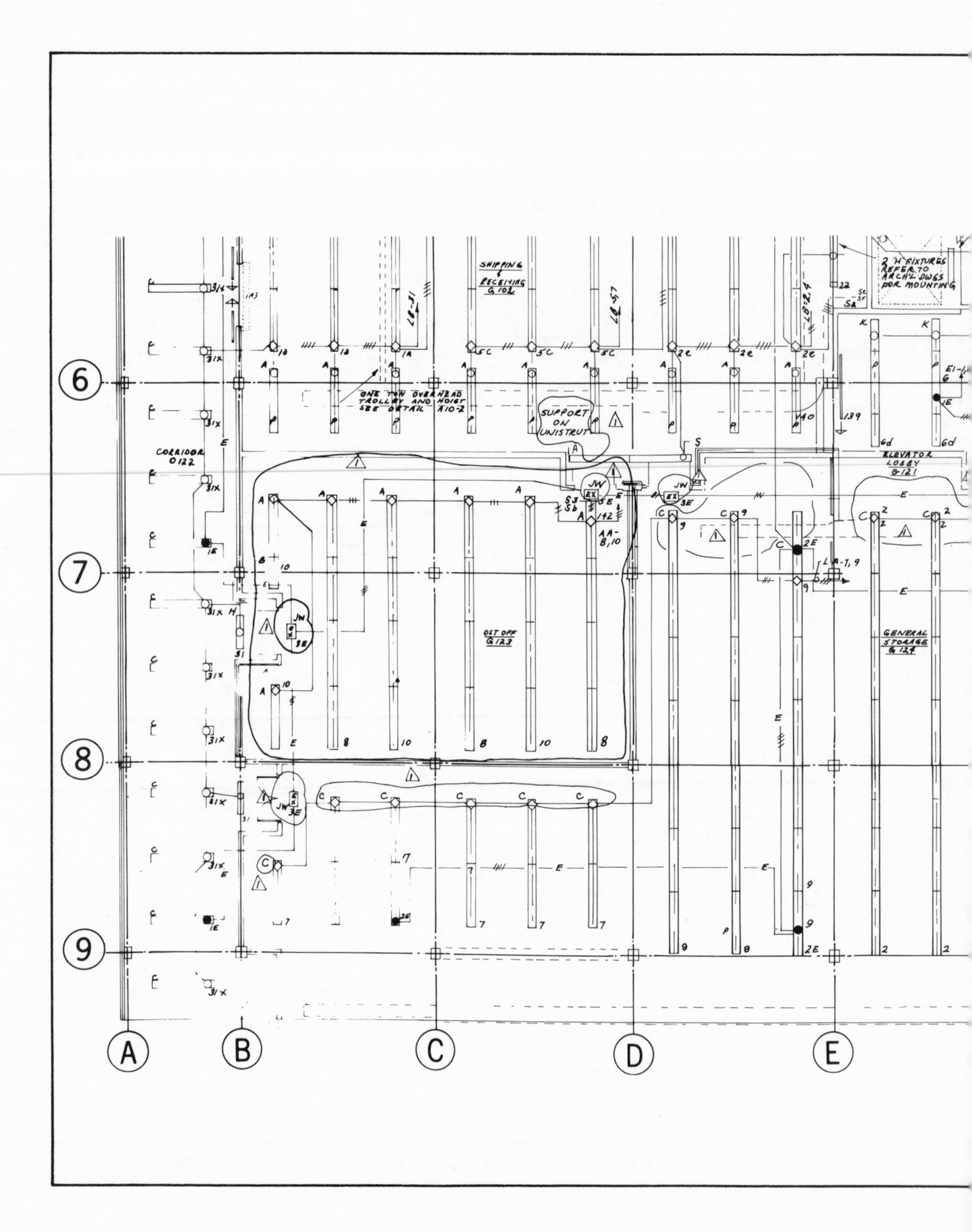

SHIPPING & RECEIVING G 102
2 'H' FIXTURES REFER TO ARCH'L DWGS FOR MOUNTING
ONE TON OVERHEAD TROLLEY AND HOIST SEE DETAIL A10-2
SUPPORT ON UNISTRUT
CORRIDOR 0122
ELEVATOR LOBBY G121
DST OFF G123
GENERAL STORAGE G124
6
7
8
9
A
B
C
D
E

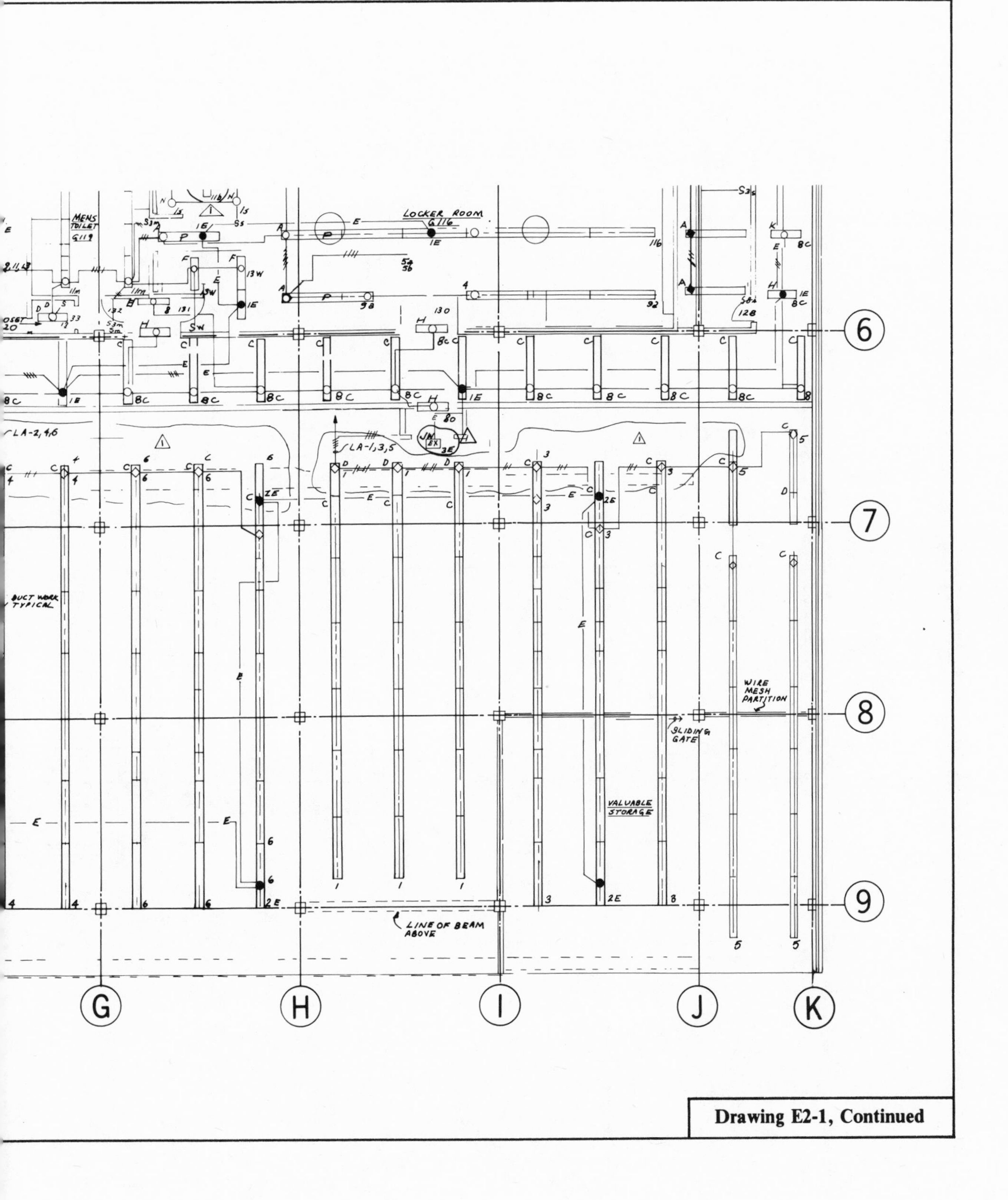

Drawing E2-1, Continued

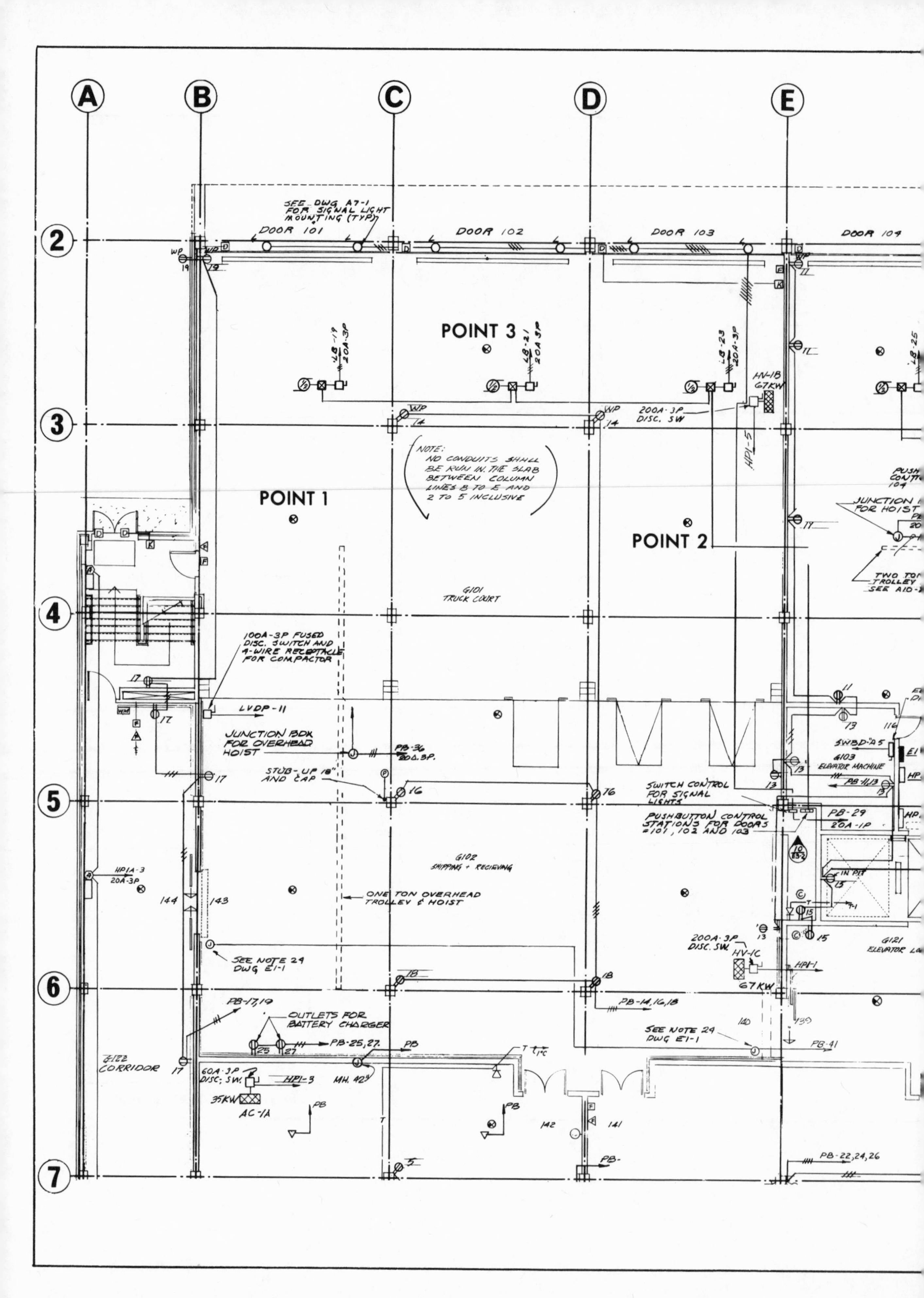

A
B
C
D
E
2
3
4
5
6
7
SEE DWG A7-1 FOR SIGNAL LIGHT MOUNTING (TYP)
DOOR 101
DOOR 102
DOOR 103
DOOR 104
POINT 3
POINT 1
POINT 2
LB-19 20A-3P
LB-21 20A-3P
LB-23 20A-3P
HV-1B 67KW
200A-3P DISC. SW
HP1-5
NOTE: NO CONDUITS SHALL BE RUN IN THE SLAB BETWEEN COLUMN LINES B TO E AND 2 TO 5 INCLUSIVE
JUNCTION BOX FOR HOIST
TWO TON TROLLEY SEE A10-
G101 TRUCK COURT
100A-3P FUSED DISC. SWITCH AND 4-WIRE RECEPTACLE FOR COMPACTOR
LVDP-11
JUNCTION BOX FOR OVERHEAD HOIST
PB-36 20A.3P.
STUB-UP 18" AND CAP
SWITCH CONTROL FOR SIGNAL LIGHTS
PUSHBUTTON CONTROL STATIONS FOR DOORS #101, 102 AND 103
SWBD-A5
G103 ELEVATOR MACHINE
PB-11,13
PB-29 20A-1P
IN PIT
HP1A-3 20A-3P
144
143
G102 SHIPPING + RECIEVING
ONE TON OVERHEAD TROLLEY & HOIST
SEE NOTE 24 DWG E1-1
200A-3P DISC. SW.
HV-1C
67KW
HP1-1
G121 ELEVATOR LO
PB-17,19
OUTLETS FOR BATTERY CHARGER
PB-25,27.
PB-14,16,18
SEE NOTE 24 DWG E1-1
PB-41
Z122 CORRIDOR
60A-3P DISC. SW.
HP1-3
35KW
AC-1A
MH. 42"
142
141
140
139
PB-22,24,26

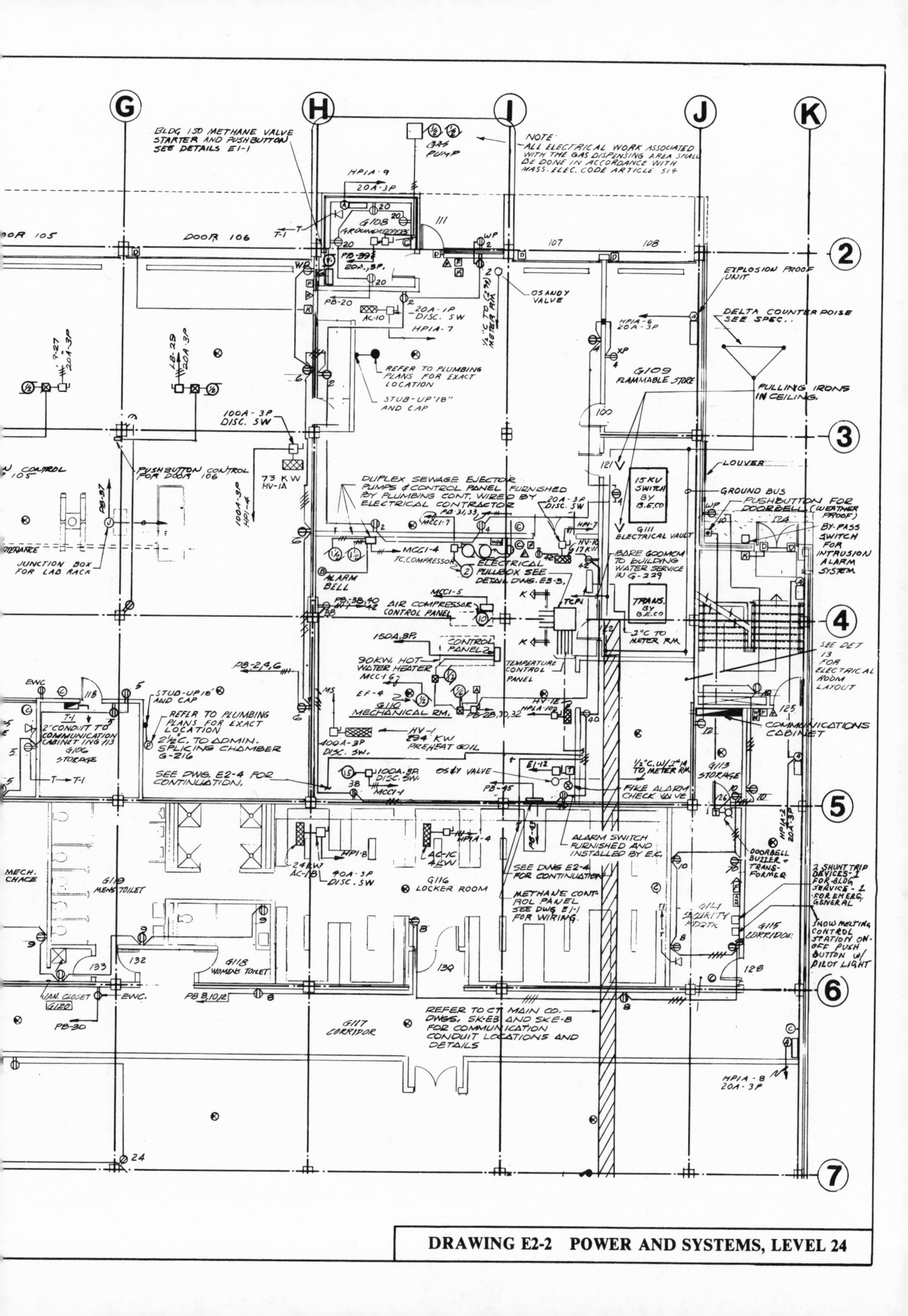

DRAWING E2-2 POWER AND SYSTEMS, LEVEL 24

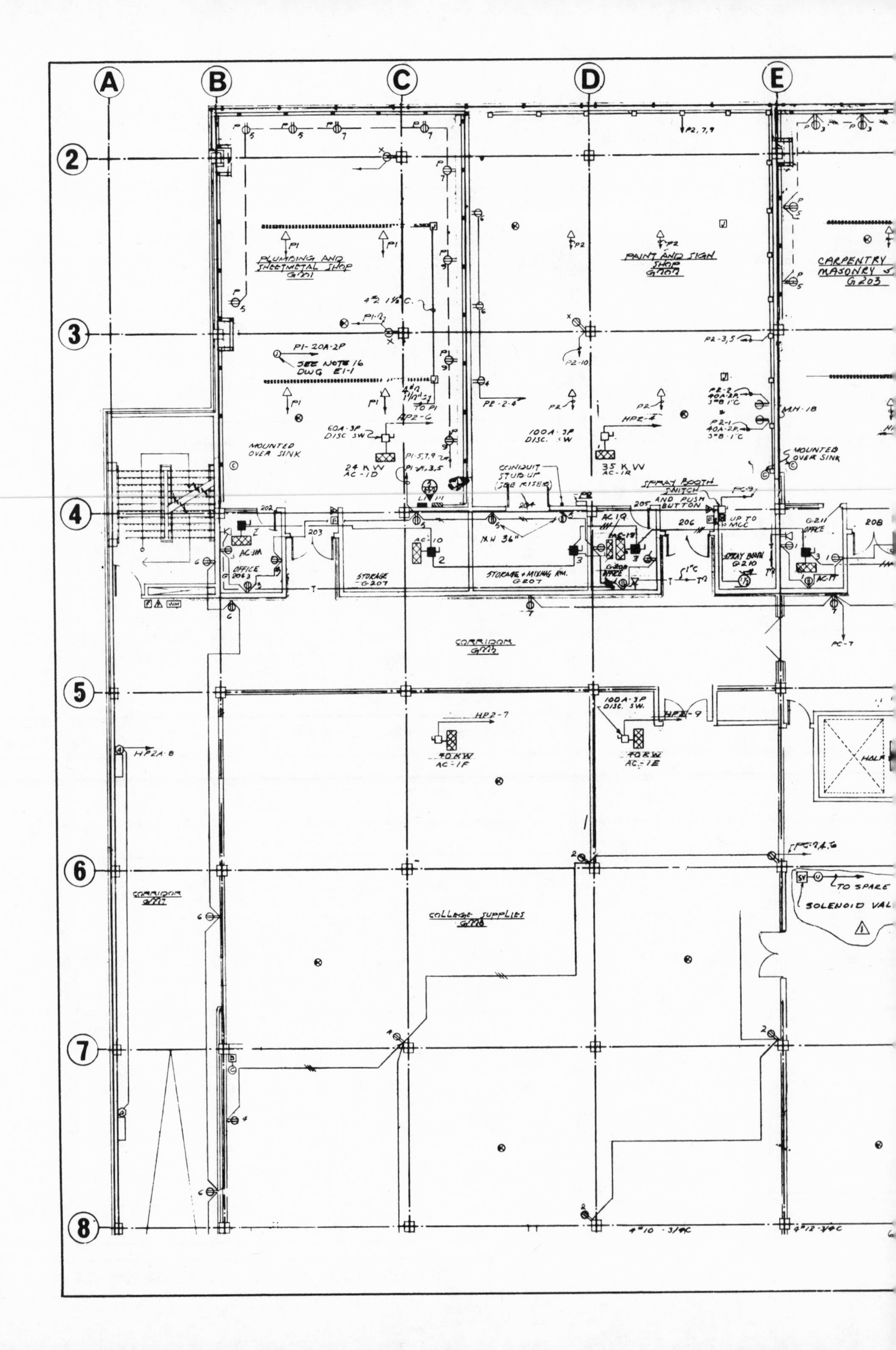

A
B
C
D
E
2
3
4
5
6
7
8
PLUMBING AND SHEETMETAL SHOP G201
PAINT AND SIGN SHOP G207
CARPENTRY MASONRY
G203
P1-20A-2P
SEE NOTE 16 DWG E1-1
60A-3P DISC SW
24 KW AC-1D
100A-3P DISC. SW
35 KW AC-1R
MOUNTED OVER SINK
CONDUIT STUB UP (SEE RISER)
SPRAY BOOTH SWITCH AND PUSH BUTTON
UP TO MCC
SPRAY BOOTH G210
STORAGE G207
STORAGE + MIXING RM. G207
OFFICE
CORRIDOR
100A-3P DISC. SW.
AC-1F
AC-1E
COLLEGE SUPPLIES
SOLENOID VAL
TO SPARE
MH 36"
MH-18

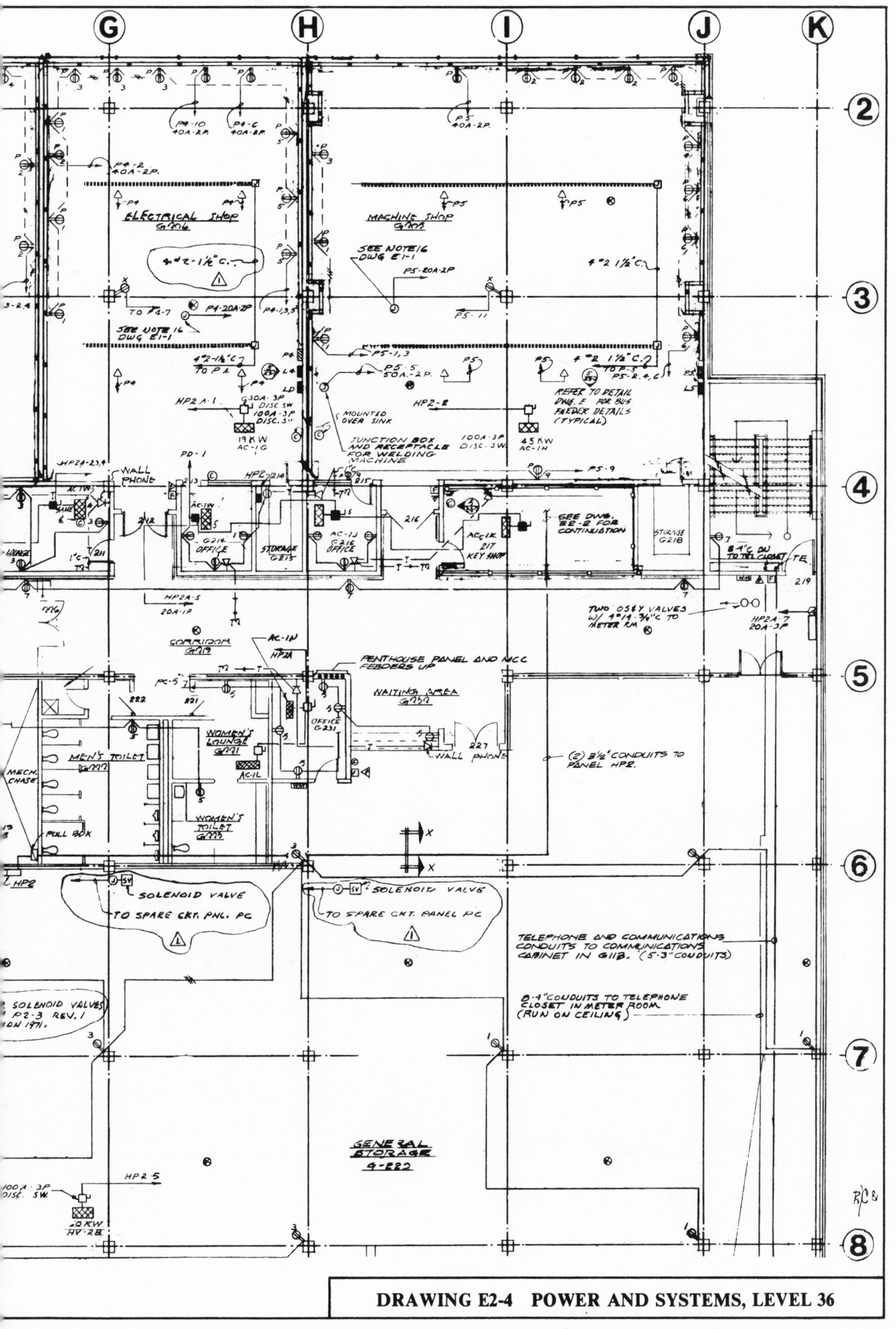

DRAWING E2-4 POWER AND SYSTEMS, LEVEL 36

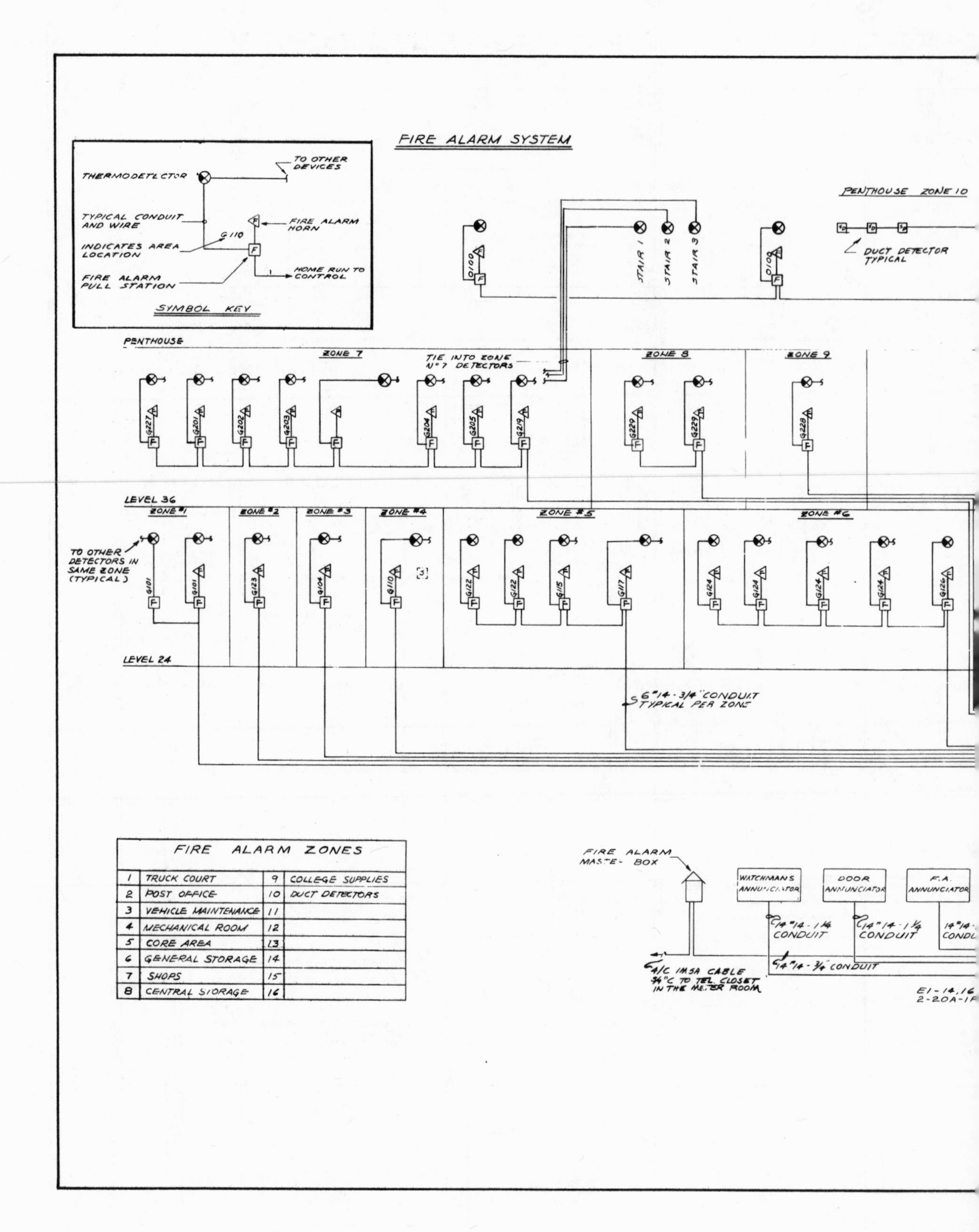

FIRE ALARM ZONES			
1	TRUCK COURT	9	COLLEGE SUPPLIES
2	POST OFFICE	10	DUCT DETECTORS
3	VEHICLE MAINTENANCE	11	
4	MECHANICAL ROOM	12	
5	CORE AREA	13	
6	GENERAL STORAGE	14	
7	SHOPS	15	
8	CENTRAL STORAGE	16	

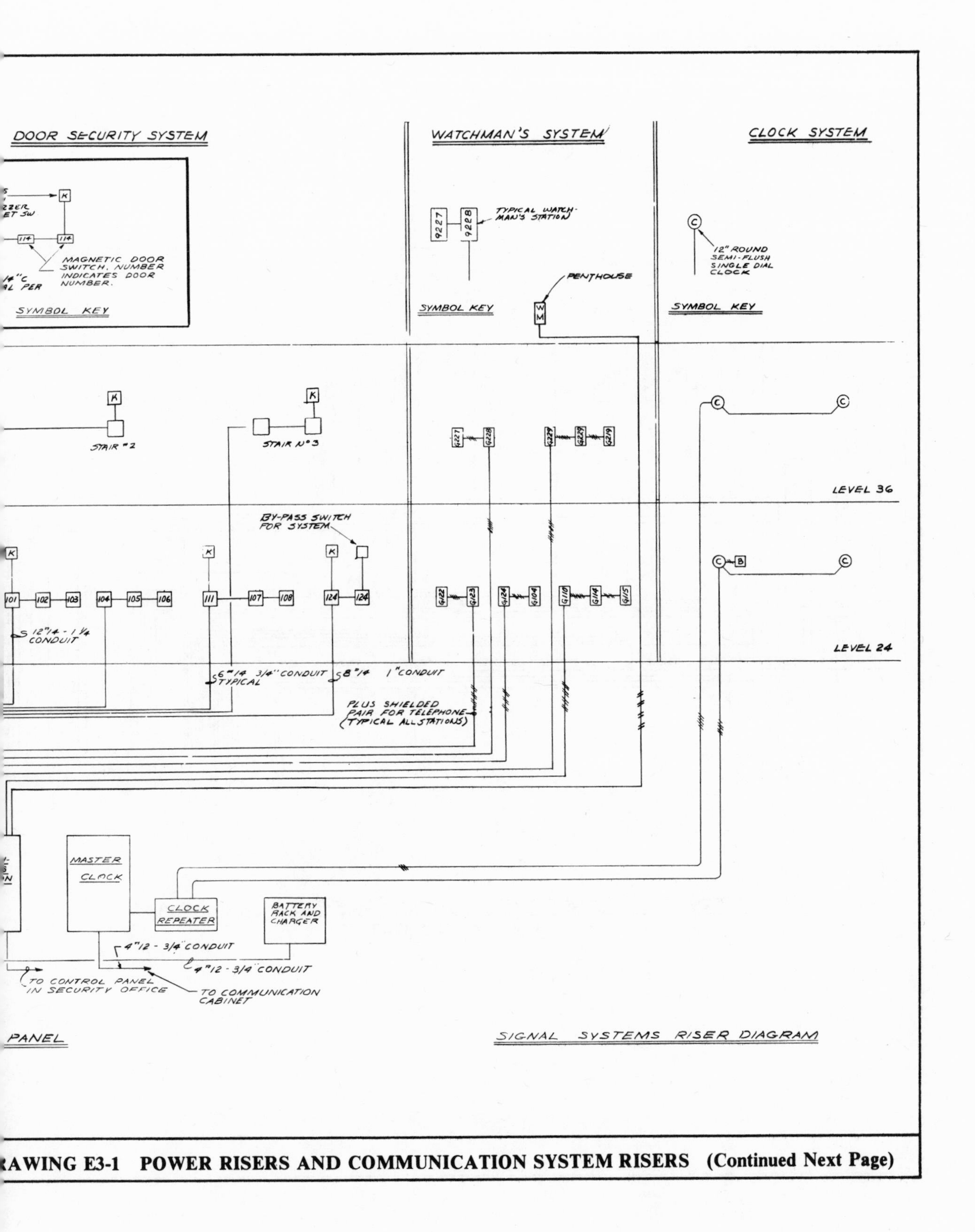

RAWING E3-1 POWER RISERS AND COMMUNICATION SYSTEM RISERS (Continued Next Page)

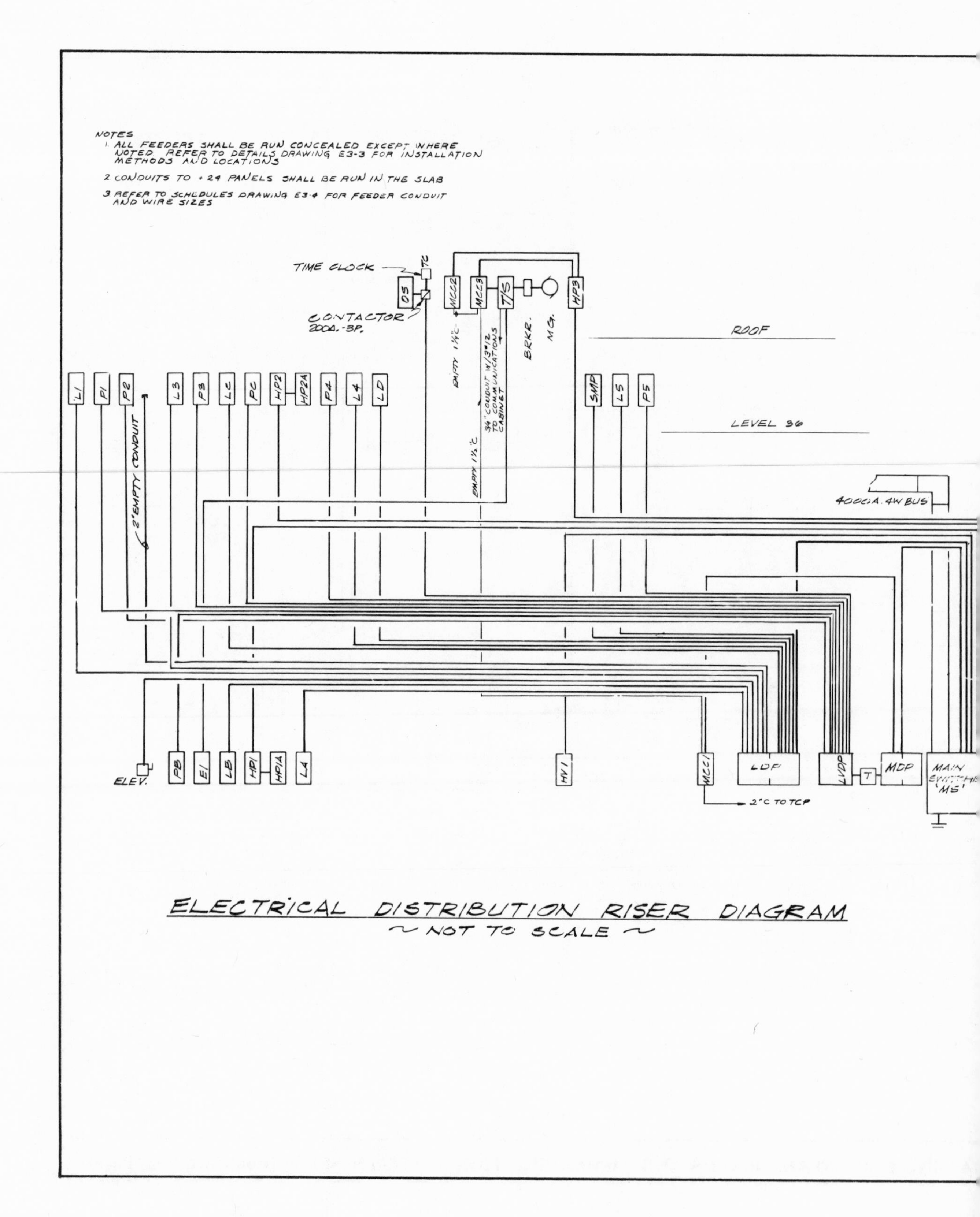
NOTES
1. ALL FEEDERS SHALL BE RUN CONCEALED EXCEPT WHERE NOTED. REFER TO DETAILS DRAWING E3-3 FOR INSTALLATION METHODS AND LOCATIONS
2. CONDUITS TO +24 PANELS SHALL BE RUN IN THE SLAB
3. REFER TO SCHEDULES DRAWING E3-4 FOR FEEDER CONDUIT AND WIRE SIZES
TIME CLOCK
TC
OS
CONTACTOR 200A.-3P.
MCC2
MCC3
T/S
HP3
EMPTY 1½"C
3/4" CONDUIT W/3#12 TO COMMUNICATIONS CABINET
BRKR.
M.G.
ROOF
L1
P1
P2
2" EMPTY CONDUIT
L3
P3
LC
PC
HP2
HP2A
P4
L4
LD
SMP
L5
P5
LEVEL 36
EMPTY 1½"C
4000A. 4W BUS
ELEV.
PB
E1
LB
HP1
HP1A
LA
HV1
MCC1
LDP
LVDP
T
MDP
MAIN SWITCHB. 'MS'
2"C TO TCP
ELECTRICAL DISTRIBUTION RISER DIAGRAM
~ NOT TO SCALE ~

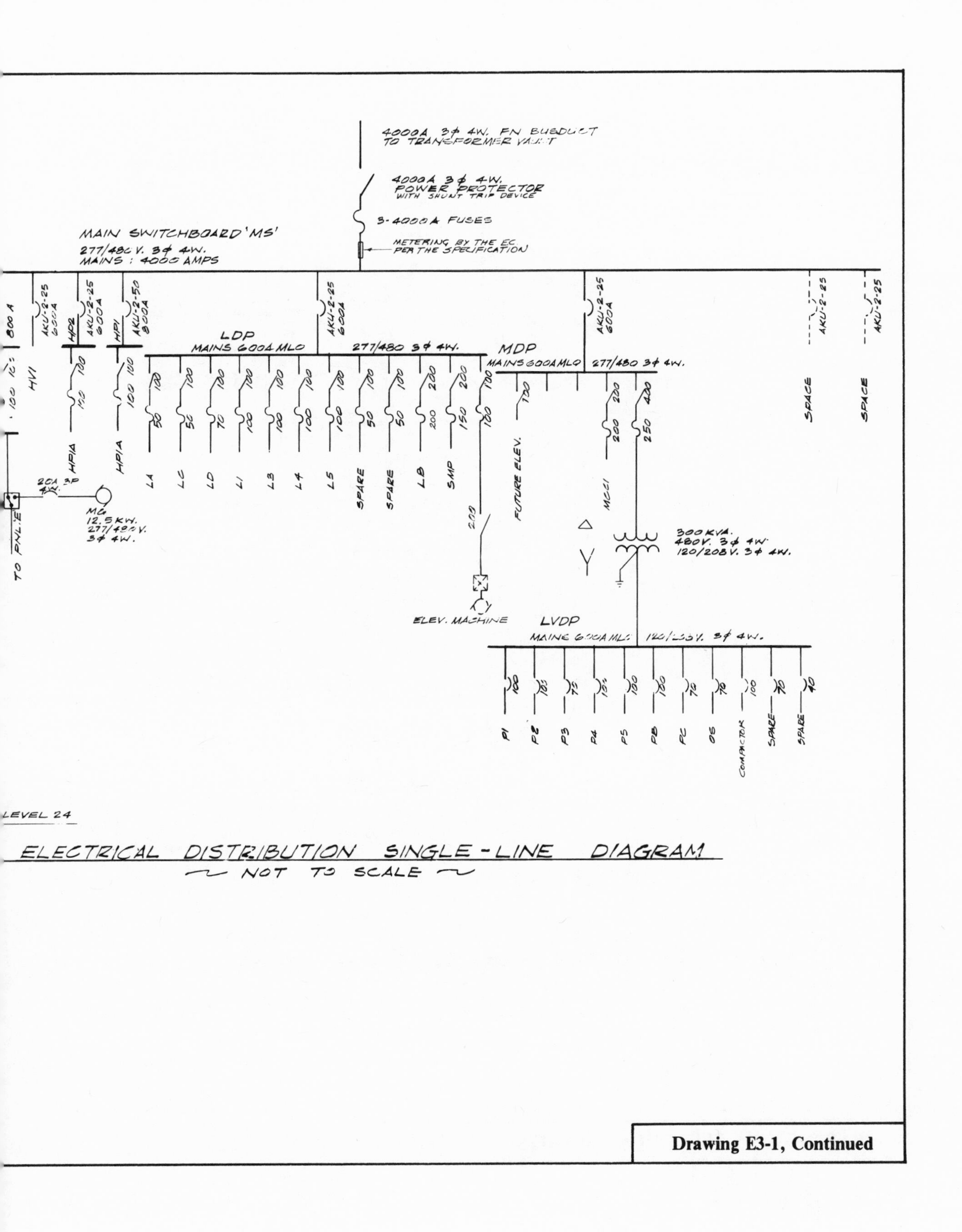

Drawing E3-1, Continued

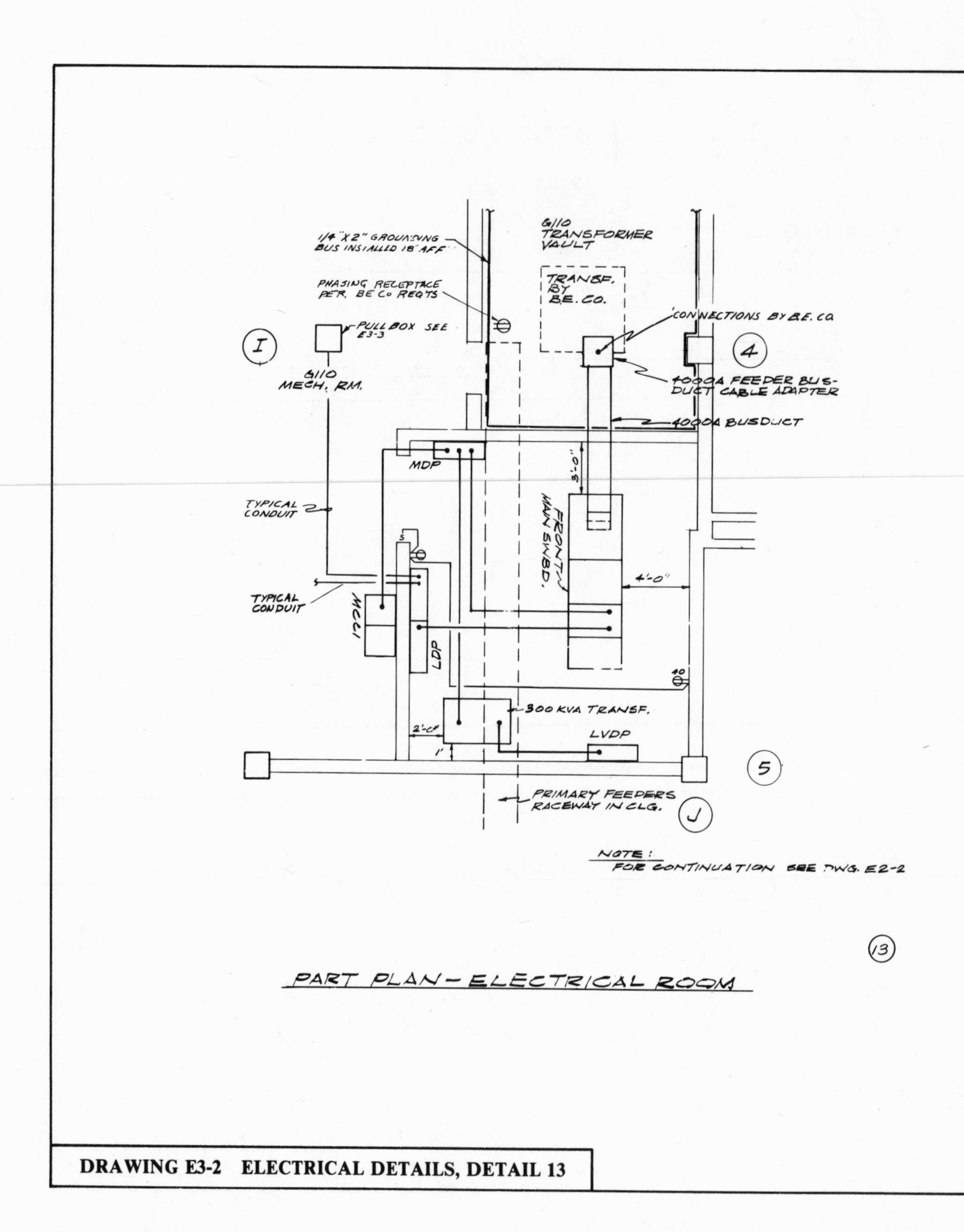

DRAWING E3-2 ELECTRICAL DETAILS, DETAIL 13

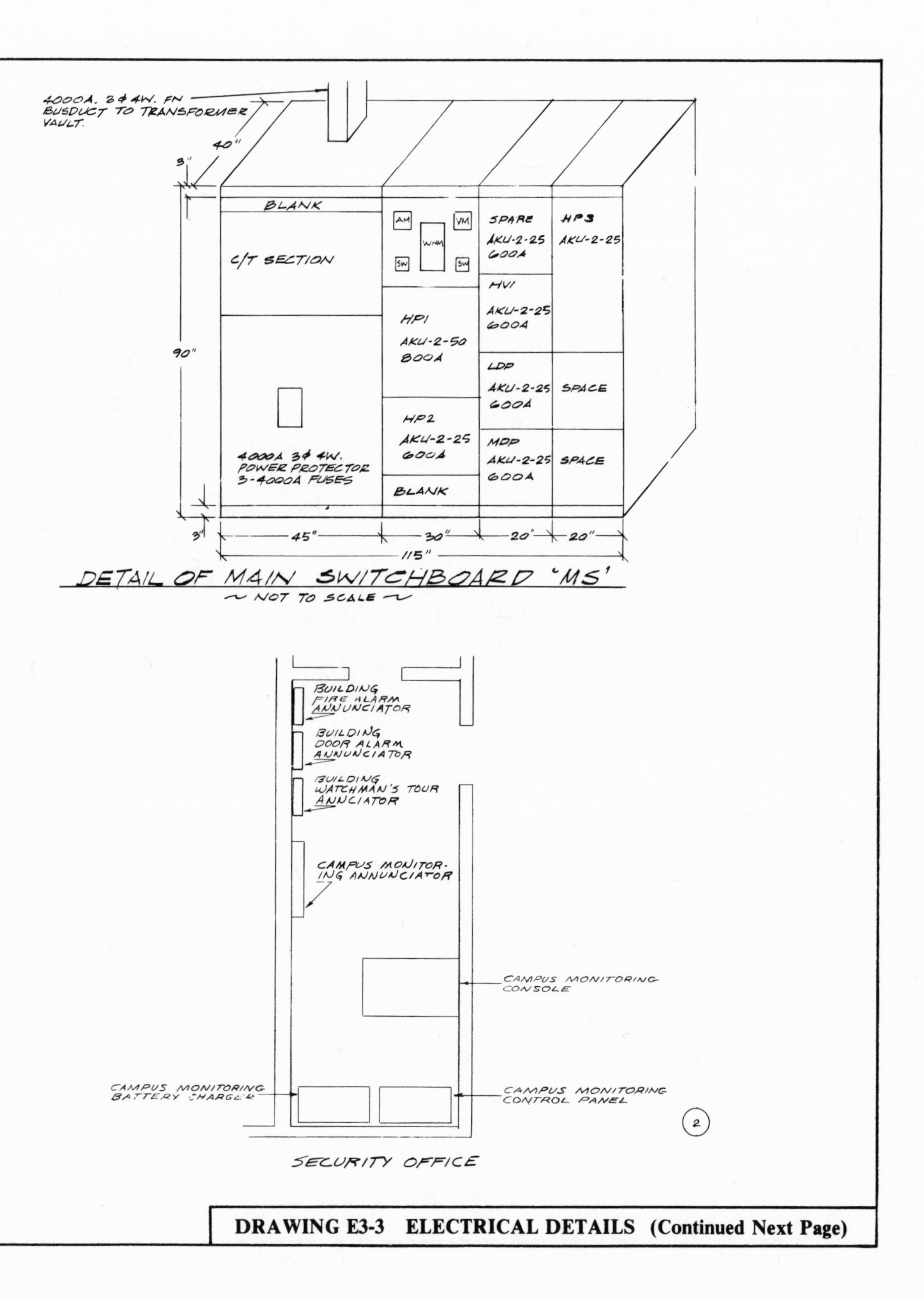

DRAWING E3-3 ELECTRICAL DETAILS **(Continued Next Page)**

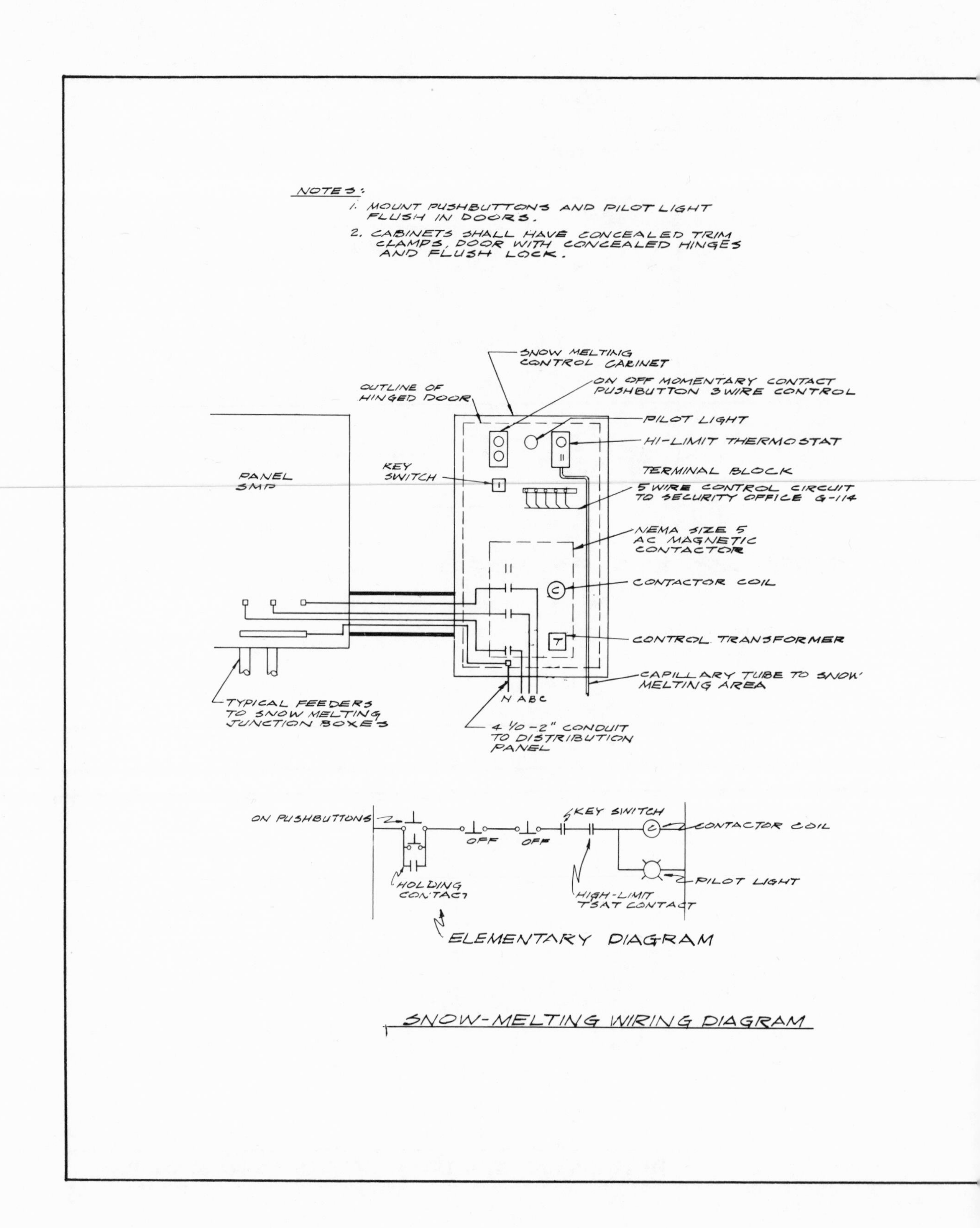
NOTES:
1. MOUNT PUSHBUTTONS AND PILOT LIGHT FLUSH IN DOORS.
2. CABINETS SHALL HAVE CONCEALED TRIM CLAMPS, DOOR WITH CONCEALED HINGES AND FLUSH LOCK.
SNOW MELTING CONTROL CABINET
ON OFF MOMENTARY CONTACT PUSHBUTTON 3 WIRE CONTROL
OUTLINE OF HINGED DOOR
PILOT LIGHT
HI-LIMIT THERMOSTAT
TERMINAL BLOCK
PANEL SMP
KEY SWITCH
5 WIRE CONTROL CIRCUIT TO SECURITY OFFICE G-114
NEMA SIZE 5 AC MAGNETIC CONTACTOR
CONTACTOR COIL
CONTROL TRANSFORMER
CAPILLARY TUBE TO SNOW MELTING AREA
N ABC
TYPICAL FEEDERS TO SNOW MELTING JUNCTION BOXES
4 1/0-2" CONDUIT TO DISTRIBUTION PANEL
ON PUSHBUTTONS
OFF
OFF
KEY SWITCH
CONTACTOR COIL
PILOT LIGHT
HOLDING CONTACT
HIGH-LIMIT TSAT CONTACT
ELEMENTARY DIAGRAM
SNOW-MELTING WIRING DIAGRAM

SNOW MAT SCHEDULE

TYPE	VOLTS	PHASE	AMPS	WATTS/ft	LENGTH	WIDTH	NOTES
A	277	1	3.1	60	6'	9"	3 SECTION MAT
~~B~~	~~277~~	~~1~~	~~4.8~~	~~44~~	~~5'~~	~~6'~~	DELETED
C	277	1	18	44	20'	6'	
D	277	1	9	44	10'	6'	
~~E~~	~~277~~	~~1~~	~~7.7~~	~~44~~	~~8'~~	~~6'~~	DELETED
F	277	1	18	44	20'	6'	W/NOTCH FOR STAIR
G	277	1	18	44	20'	6'	W/NOTCH FOR STAIR

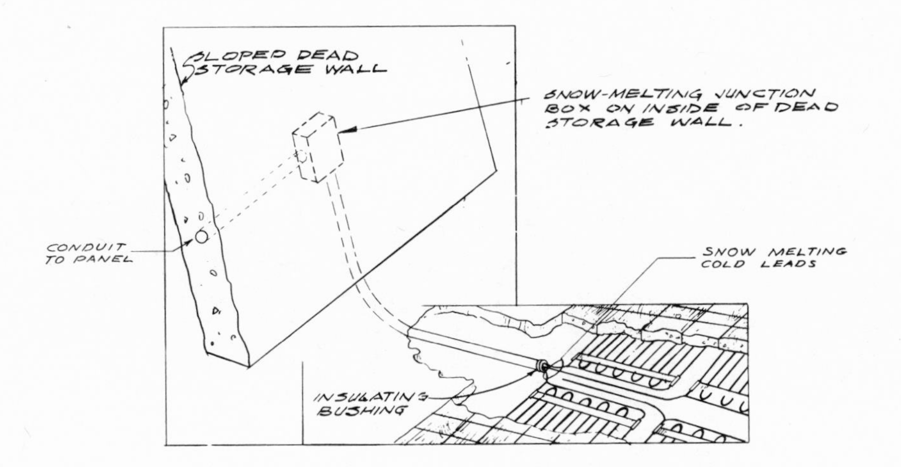

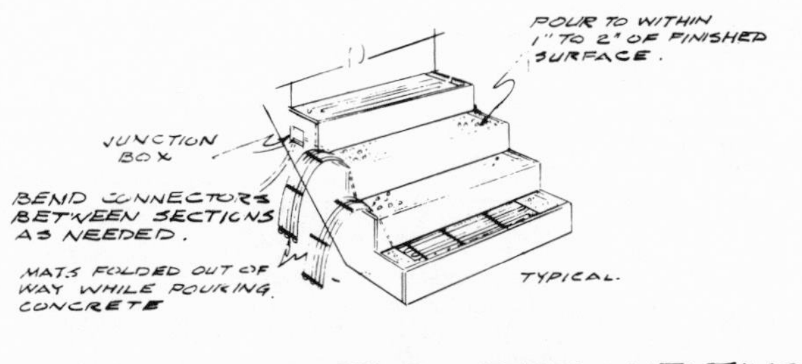

STAIR - SNOW MELTING INSTALLATION

Drawing E3-3, Continued

SCHEDULE MAIN SWITCHBOARD 'MS'

277/480V, 3φ, 4W.

MAINS: 4000 AMPS FUSIBLE POWER PROTECTOR
MAIN DEVICE: GE TYPE LB-1
4000A 277/480V 3φ 4W
FUSIBLE POWER PROTECTOR
3-4000A FUSES

MAIN FEEDER:
4000A 3φ 4W
FN BUSDUCT

NO.	SERVES	BREAKERS			FEEDERS		REMARKS
		TRIP	POLE	FRAME	CONDUCTOR	CONDUIT	
1	DIST. PNL. 'HP1'	800	3	AKU-2-50	(3) 4-250 MCM	(3) 3"	⑤
2	DIST. PNL. 'HP2'	600	3	AKU-2-25	(2) 4-350 MCM	(2) 3½"	
3	DIST. PNL. 'HP3'	800	3	AKU-2-50	(2) 4-500 MCM	(2) 4"	
4	DIST. PNL. 'LDP'	600	3	AKU-2-25	(2) 4-350 MCM	(2) 3½	
5	DIST. PNL. 'MDP'	600	3	AKU-2-25	(2) 4-350 MCM	(2) 3½"	
6	HTG. UNIT 'HV1'	600	3	AKU-2-25	(2) 4-350 MCM	(2) 3½"	
7	SPACE FOR AKU-2-25	—	—	—	—	—	

SCHEDULE DISTRIBUTION PANEL LDP

277/480V 3φ 4W.

MAINS: 600A MLO
DOUBLE SECTION PANEL

SURFACE MOUNTED
TYPE QMR

FEEDER: (2) 4-350 MCM
3½" C./EA.

CCT. NO.	SERVES	SWITCH			FEEDER		REMARKS
		SWITCH	FUSE	POLE	CONDUCTOR	CONDUIT	
1	PANEL 'LA'	60	40	3	4 #8	1½"	
2	PANEL 'LB'	200	200	3	4 # 3/0	2½"	⑤
3	PANEL 'LC'	60	50	3	4 #6	1¼"	
4	PANEL 'LD'	100	70	3	4 #4	1½"	
5	PANEL 'L1'	100	100	3	4 #2	2"	
6	PANEL 'L3'	100	100	3	4 #2	2"	
7	PANEL 'L4'	100	100	3	4 #2	2"	
8	PANEL 'L5'	100	100	3	4 #2	2"	
9	PANEL 'SMP'	200	150	3			
10	ELEVATOR	200	100	3	4 #2	2"	
11	SPARE	60	40	3	—	—	
12	SPARE	60	50	3	—	—	
13	SPACE	100	—	3			
14	SPACE	100	—	3			

SCHEDULE DISTRIBUTION PANEL MDP

277/480V. 3φ 4W.

MAINS: 600A. MLO

SURFACE MOUNTED
TYPE QMR

FEEDER: (2) 4-350 MCM
3½" C./EA.

CCT. NO.	SERVES	SWITCH			FEEDER		REMARKS
		SWITCH	FUSE	POLE	CONDUCTOR	CONDUIT	
1	FUT. ELEV. MACH.	200	100	3	4 #2	2"	
2	MCC1	200	200	3	3 # 3/0	2½"	
3	TRANSFORMER (LVDP)	400	250	3	3-250 MCM	3"	
4	SPACE	200	—	3			
5	SPACE	100	—	3			

MOTOR CONTROL CENTER MCC1
480 VOLT 3 PHASE 3 WIRE

MAINS: 200A. MLO
NEMA CLASS I, TYPE "B"

FEEDER: 3#4/0
CONDUIT 3"C.

NO.	SERVES	H.P.	POLE	STARTER SIZE	SWITCH SIZE	FUSE SIZE	FEEDER		INTERLOCKS WITH
							COND.	CONDUIT	
1	H&V-1	15	3	2	30	25	3#10	3/4"	E-2, E-3, E-4, LIMIT SW. SMOKE DETECTOR
2	SPARE	7½	3	1	30	15	3#12	3/4"	—
3	SPARE	7½	3	1	30	15	3#12	3/4"	—
4	DUPLEX EJECTOR	2-1½	3	NONE	30	15	3#12	3/4"	LIMIT SWITCH
5	COMPRESSOR	10	3	NONE	60	40	3#12	3/4"	P.E.
6	EF 4	½	3	00	30	1⅛	3#12	3/4"	H&V-1
7	METER RM	—	3	1	60	60	3#4	1½	
8	SPACE	—	3	2	—	—	—	—	
9	SPARE			1	30	—	—	—	

MOTOR CONTROL CENTER MCC2
480 VOLT 3 PHASE 3 WIRE

MAINS: 200A MLO
NEMA CLASS I, TYPE "B"

FEEDER: 3#3/0
CONDUIT 2½"C.

NO.	SERVES	H.P.	POLE	STARTER SIZE	SWITCH SIZE	FUSE SIZE	FEEDER		INTERLOCKS WITH
							COND.	CONDUIT	
1	AC1	30	3	3	60	45	3#6	1½"	FREEZE PROT'N T'STAT SMOKE DETECTOR
2	HV2	10	3	1	30	17½	3#12	3/4"	HI-LIMIT SWITCH
3	SPARE	10	3	1	30	17½	3#12	3/4"	—
4	HV3	3	3	0	30	5 6/10	3#12	3/4"	SPRAY BOOTH FAN
5	CP1	7½	3	1	30	15	3#12	3/4"	
6	CP2	7½	3	1	30	15	3#12	3/4"	
7	SPARE		3	1	—	—	—	—	
8	SPARE		3	3	—	—	—	—	

MOTOR CONTROL CENTER MCC3
480 VOLT 3 PHASE 3 WIRE

MAINS: 100A. MLO
NEMA CLASS I, TYPE "B"

FEEDER: 3#1
CONDUIT 2"C.

NO.	SERVES	H.P.	POLE	STARTER SIZE	SWITCH SIZE	FUSE SIZE	FEEDER		INTERLOCKS WITH
							COND.	CONDUIT	
1	E1 (FAN)	1½	3	00	30	2 8/10	3#12	3/4"	SMOKE DETECTOR
2	E2 (FAN)	10	3	1	30	17½	3#12	3/4"	SMOKE DETECTOR H&V-1
3	E3 (FAN)	2	3	00	30	5 6/10	3#12	3/4"	H&V-1
4	SPARE	5	3	0	30	9	3#12	3/4"	
5	EMERG. GEN. 'MG'		3	NONE	30	20	4#12	3/4"	
6	SPACE		3	1	—	—			
7	SPACE		3	1	—	—			

DRAWING E3-4 SWITCHBOARD AND PANEL SCHEDULES **(Continued Next Page)**

SCHEDULE DISTRIBUTION PANEL HP1

277/480V 3φ 4W

MAINS: 800A MLO — DOUBLE SECTION PANEL

SURFACE MOUNTED — TYPE QMR

FEEDER: (3) 4-250MCM 3" C./EA.

CCT. NO.	SERVES	SWITCH			FEEDERS		REMARKS
		SWITCH	FUSE	POLES	CONDUCTOR	CONDUIT	
1	REHEAT HV-1C	200	125	3	3#1	2"	
2	SPARE	60	40	3	3#8	1"	
3	REHEAT A[illegible]-1A	100	70	3	3#4	1½"	
4	REHEAT HV-1A	200	125	3	3#1	2"	
5	REHEAT HV-1B	200	125	3	3#1	2"	
6	REHEAT HV-2A	200	125	3	3#1	2"	
7	REHEAT HV-10	30	30	3	3#8	1"	
8	REHEAT AC-1B	60	40	3	3#8	1"	
9	HP1A	100	100	3	4#2	2"	
10	SPARE	100	70	3	—	—	
11	SPACE	30	—	—	—	—	
12	SPACE	60	—	—	—	—	

SCHEDULE DISTRIBUTION PANEL HP1A

277/480V. 3φ 4W.

MAINS: 100A MLO

SURFACE MOUNTED — TYPE

FEEDER: 4#2 2" C.

CCT. NO.	SERVES	BREAKERS			FEEDER		REMARKS
		TRIP	POLE	FRAME	CONDUCTOR	CONDUIT	
1	CUH	20	3	[illegible]	3#12	¾"	
2	CUH	20	3	[illegible]	3#12	¾"	
3	CUH	20	3	100	3#12	¾"	
4	AC1C	20	3	100	3#12	¾"	
5	CUH	20	3	100	3#12	¾"	
6	CUH	20	3	100	3#12	¾"	
7	AC 1[illegible]	[illegible]	1	100	2#12	½	
8	CUH	[illegible]	3	100	3#12	¾	
9	CUH	[illegible]	3	100	3#12	¾"	
10	HV1E	20	3	100	3#12	¾"	
11	AC1N + AC1L	20	3	100	2#12	½"	
12-15	SPARE	20	3	100	—	—	

SCHEDULE DISTRIBUTION PANEL HP2

277/480V 3φ 4W.

MAINS: 600A MLO

SURFACE MOUNTED — TYPE QMR

FEEDER: (2) 4-350MCM 3½" C./EA.

CCT. NO.	SERVES	SWITCH			FEEDER		REMARKS
		SWITCH	FUSE	POLE	CONDUCTOR	CONDUIT	
1	SPARE	100	70	3	3#4	1½"	
2	REHEAT AC-1H	100	70	3	3#4	1½"	
3	REHEAT AC-1V	100	70	3	3#4	1½"	
4	REHEAT AC-1R	100	70	3	3#4	1½"	
5	REHEAT HV-2B	100	100	3	4#2	2"	
6	REHEAT AC-10	100	70	3	3#4	1½"	
7	REHEAT AC-1F	100	70	3	3#4	1½"	
8	HP2A	100	100	3	4#2	2"	
9	REHEAT AC-1E	100	70	3	3#4	1½"	
10	SPACE	60	—	3	—	—	
11	SPACE	60	—	3	—	—	
12	SPACE	100	—	3	—	—	

SCHEDULE DISTRIBUTION PANEL HP2A

277/480V. 3ϕ 4W.

MAINS: 100A. MLO — SURFACE MOUNTED TYPE — FEEDER: 4#2 2" C.

CCT. NO.	SERVES	BREAKERS			FEEDER		REMARKS
		TRIP	POLE	FRAME	CONDUCTOR	CONDUIT	
1	REHEAT	30	3	100	3#10	3/4"	
2	REHEAT	20	1	100	2#12	1/2"	
3	REHEAT	20	1	100	2#12	1/2"	
4	REHEAT	20	1	100	2#12	1/2"	
5	REHEAT	20	1	100	2#12	1/2"	
6	SPARE	20	3	100	3#12	3/4"	
7	CUH	20	3	100	—	—	
8	2 CUH	20	3	100	3#12	3/4"	
9	SPARE	20	1	100	—	—	
10-12	SPACE	20	3	100	—	—	

SCHEDULE DISTRIBUTION PANEL HP3

277/480V. 3ϕ 4W.

MAINS: 800A MLO — SURFACE MOUNTED TYPE QMR — FEEDER: (2) 4-350MCM 3½" C./EA.

CCT. NO.	SERVES	SWITCH			FEEDER		REMARKS
		SWITCH	FUSE	POLE	CONDUCTOR	CONDUIT	
1	HV UNIT HV-3	400	300	3	3-300 MCM	8"	
2	HV UNIT HV-3A	100	70	3	3#4	1½"	
3	UNIT HEATER UH-1	60	50	3	3#6	1¼"	
4	UNIT HEATER UH-1	60	50	3	3#6	1¼"	
5	MCC-2	200	200	3	3#4/0	3"	
6	MCC-3	100	100	3	3#1	2"	
7	METHANE CB PANEL	60	60	3	4"6	1½"C	
8	SPARE	100	—	3			
9-12	SPACE	60	—	3			

SCHEDULE DISTRIBUTION PANEL LVDP

120/208V. 3ϕ 4W.

MAINS: 600A. MLO — SURFACE MOUNTED TYPE CCB — FEEDER: (2) 4-350MCM 3½" C./EA.

CCT. NO.	SERVES	BREAKERS			FEEDER		REMARKS
		TRIP	POLE	FRAME	CONDUCTOR	CONDUIT	
1	SPARE	70	3	100	—	—	
2	PANEL 'P1'	100	3	100	4#2	2"	
3	PANEL "P2"	70	3	100	—	—	
4	PANEL 'P3'	70	3	100	4#4	1½"	
5	PANEL 'P4'	100	3	100	4#2	2"	
6	PANEL 'P5'	100	3	100	4#2	2"	
7	SPARE	40	3	100	—	—	
8	PANEL 'PE'	100	3	100	4#2	2"	⑤
9	PANEL 'PC'	70	3	100	4#4	1½"	
10	PANEL 'OS'	70	3	100	4#4	1½"	
11	COMPACTOR	100	3	100	4#2	2"	
12	SPACE	100		100			

Drawing E3-4, Continued

277/480 VOLT PANELBOARD SCHEDULE

BRANCH BREAKER : 100A FRAM (BOLT IN TYPE)

BUSSING : AS INDICATED

MARK	MOUNTING	MIN. MAIN BUS	FLOOR	LOCATION	BREAKERS							REMARKS
					20A-1P	20A-2P	20A-3P	30A-1P	30A-2P			
L1	SURF	100	+36	G-201	1	2	2		1			①
L2	SURF	100	+36	FUTURE	—	—	—	—	—			EMPTY CONDUIT
L3	SURF	100	+36	G-203	1	2	2		1			①
L4	SURF	100	+36	G-204	1	2	2		1			①
L5	SURF	100	+36	G-205	1	2	2		1			①
L6	SURF	—	—	DELETED	—	—	—	—	—			
L7	SURF	—	—	↓								
L8	SURF	—	—	↓								
L9	SURF	—	—	↓								
L10	SURF	—	—	↓								
E1	SURF	100	+24	G-104	24							
SMP	SURF	225	+36	G-229	10			20				
LA	SURF	100	+24	G-124	16							
LB	SURF	225	+24	G-104	27		8					④
LC	SURF	100	+36	G-229	18							
LD	SURF	100	+36	G-204	24							

120/208 VOLT PANELBOARD SCHEDULE

BRANCH BREAKERS 100A FRAME (BOLT IN TYPE)

BUSSING : AS INDICATED

MARK	MOUNTING	MIN. MAIN BUS	FLOOR	LOCATION	BREAKERS							REMARKS
					20A-1P	20A-2P	20A-3P	30A-1P	30A-2P	40A 2P	50A 2P	
P1	SURF	100	+36	G-201	10	2	1	1	1			②
P2	SURF	100	+36	G-202	14	2	1	1	1			②
P3	SURF	100	+36	G-203	10							②
P4	SURF	100	+36	G-204	8	2	1	1	1	3		②
P5	SURF	100	+36	G-205	10	2	1	1	1	1	1	②
P6	SURF	100		DELETED								
PA	SURF	100		DELETED								
PB	SURF	225	+24	G-104	45	4	4		4			④
PC	SURF	225	+36	G-229	22	2	1					
OS	SURF	225	+49	PENTHOUSE	36							

REMARKS

① BUSSED FOR 24 CIRCUIT

② BUSSED FOR 36 CIRCUIT

③ BUSSED FOR 42 CIRCUIT

④ DOUBLE PANEL

⑤ RUN FEEDER IN SLAB

Drawing E3-4, Continued

APPENDIX D

TAKE-OFF SHEETS, PRICE SHEETS, AND SUMMARY SHEET

TABLE TO-1 LIGHTING FIXTURES (Part 1 of 2)

LOCATION

ARCHITECT ENGINEER

CLASSIFICATION *LIGHTING FIXTURES*

DESCRIPTION	NO.	DIMENSIONS			A	B	C	D
E2-1					164	37	153	30
					4	1	6	1
					2		1	
					1			
E2-2								
E2-3					139		241	4
							1	
E2-5						28		
				TOTAL	310	66	409	35

TABLE TO-1 LIGHTING FIXTURES (Part 2 of 2)

LOCATION

ARCHITECT ENGINEER

CLASSIFICATION *LIGHTING FIXTURES (continued)*

DESCRIPTION	NO.	DIMENSIONS			ELEC VAULT	PLT LIGHTS	P	
E2-1					6	2		
E2-5							4	
				TOTAL	6	2	4	

TOR | ESTIMATE NO.
SIONS | SHEET NO.
ED | DATE

	H	J	K	L	M	N	EXP PROOF	EXIT	ESTIMATED QUANTITY	UNIT
2	10	1	21			4	2	5		
	1					1		5		
	2							2		
				6						
1	5		10					7		
	10									
	1									
			10		104			2		
					170					
3	29	1	41	6	274	5	2	21		

TOR | ESTIMATE NO.
SIONS | SHEET NO.
ED | DATE

									ESTIMATED QUANTITY	UNIT

TABLE TO-2 DEVICES (Part 1 of 2)

LOCATION

ARCHITECT
ENGINEER

CLASSIFICATION DEVICES

DESCRIPTION	NO.	DIMENSIONS			S1	S2	S3	S4	S
E2 -1					27		4		
							2		
E2-2									
E2-3					37		2		
E2-4									
E2-5					6				
				TOTAL	70		8		

MATOR | ESTIMATE NO.

ENSIONS | SHEET NO.

CKED | DATE

	F	F	⊗	●	F	S	H	ANN	ESTIMATED QUANTITY	UNIT
	10	8	1	36	1			1		
	2	2								
	12	10	3	41			2			
	2			8		3				
	26	20	4	85	1	3	2	1		

TABLE TO-2 DEVICES (Part 2 of 2)

LOCATION

ARCHITECT ENGINEER

CLASSIFICATION DEVICES

DESCRIPTION	NO.	DIMENSIONS			D	K	WF	wm	C
E2-2					26	6		5	
E2-4					12	6	1	5	
E2-5					4	2	1		
				TOTAL	42	14	2	10	

R

NS

SNOW MELTING MATS

ESTIMATE NO.

SHEET NO.

DATE

	C	D	F	G					ESTIMATED QUANTITY	UNIT
	11	5	1	1						
8										
8	11	5	1	1						

TABLE TO-3 SPECIAL EQUIPMENT

LOCATION

ARCHITECT ENGINEER SPECIAL EQUIPMENT -

CLASSIFICATION ELECTRIC HEAT - (HV, ETC. UNITS)

DESCRIPTION	NO.	DIMENSIONS			KW	VOLTS	PHASE	1ϕDISC 30F
E2-2	AC	1A			35	480	3	
E2-2	HV	1C			67	480	3	
E2-2	HV	1B			67	480	3	
E2-2	HV	2A			77	480	3	
E2-2	HV	1A			73	480	3	
E2-2	AC	1O			6	277	1	1
E2-2	HV	1			294	480	3	
E2-2	HP	1A			12	480	3	
E2-2	AC	1C			4	480	3	
E2-2	HV	1O			17	480	3	
E2-4	AC	1D			24	480	3	
E2-4	AC	11A			8	277	1	
E2-4	AC	1O			8	277	1	
E2-4	AC	1Q			8	277	1	
E2-4	AC	1S			8	277	1	
E2-4	AC	1T			8	277	1	
E2-4	AC	1U			8	277	1	
E2-4	AC	1W			8	277	1	
E2-4	AC	1L			8	277	1	
E2-4	AC	1J			8	277	1	
E2-4	AC	1K			8	277	1	
E2-4	AC	1F			40	480	3	
E2-4	AC	1R			35	480	3	
E2-4	AC	1E			40	480	3	
E2-4	AC	1Y			38	480	3	
E2-4	HV	2B			60	480	3	
E2-4	AC	1G			19	480	3	
E2-4	AC	1H			45	480	3	
E2-4	HV	2			185	480	3	
E2-4	HUH	2			8	277	1	
		TOTAL	TABLE	TO-4				2
		TOTAL	TABLE	TO-3				1
				TOTALS				3
HEATERS TO CONNECT								
1-30	KW		17					
35-60	KW		7					
65-100	KW		4					
101-185	KW		1					
186-300	KW		1					

TCH

ESTIMATE NO.

SHEET NO.

DATE

200F	400F	THERM	■┘					ESTIMATED QUANTITY	UNIT
1									
1									
1									
	1								
			1						
			1						
			1						
			1						
			1						
			1						
			1						
			1						
			1						
			1						
	1								
			1						
3	2		11						
3	2		11						

TABLE TO-4 MOTOR CONNECTIONS AND EQUIPMENT

LOCATION

ARCHITECT ENGINEER

CLASSIFICATION MOTOR CONN. & EQUIP

	DESCRIPTION	UNIT	HORSE-POWER HP	VOLTAGE V	PHASE Φ	MOTOR 1Φ MS	SWITCH 1 Φ 30F	3 Φ 30F
E2-2	HOT WATER HEATER		1/12	120	1			
E2-2	HOT WATER HEATER		1/6	120	1			1
E2-2	HV-1		15	480	3			
E2-2	AIR COMPRESSOR		10	480	3			
E2-2	DUPLEX SEWER EJECTOR (2)		1 1/2	480	3			
E2-2	TEMP CONTROL COMPRESSOR(2)		2	480	3			
E2-2	GASOLINE PUMPS (2)		1/2	208	1		2	
E2-2	DOOR OPENERS (O'HEAD) (6)		1/2	480	3			
E2-5	E-3		2	480	3			
E2-5	E-1		1 1/2	480	3			
	E-2		10	480	3			
	EMERGENCY GEN.	KW	12 1/2	480	3			2
	AC-1		30	480	3			
	HVB		3	480	3			
	CP 1 + 2 (2)		7 1/2	480	3			
E2-4	SPRAYBOOTH EXHAUST FAN		1 1/2	480	3			
E2-4	O S + Y VALVES (2)		1/2		1			
	TOTALS						2	3
MOTORS TO CONNECT	FRAC. TO 1/2		4					
	3/4 TO 7 1/2		7					
	10 - UP		5					

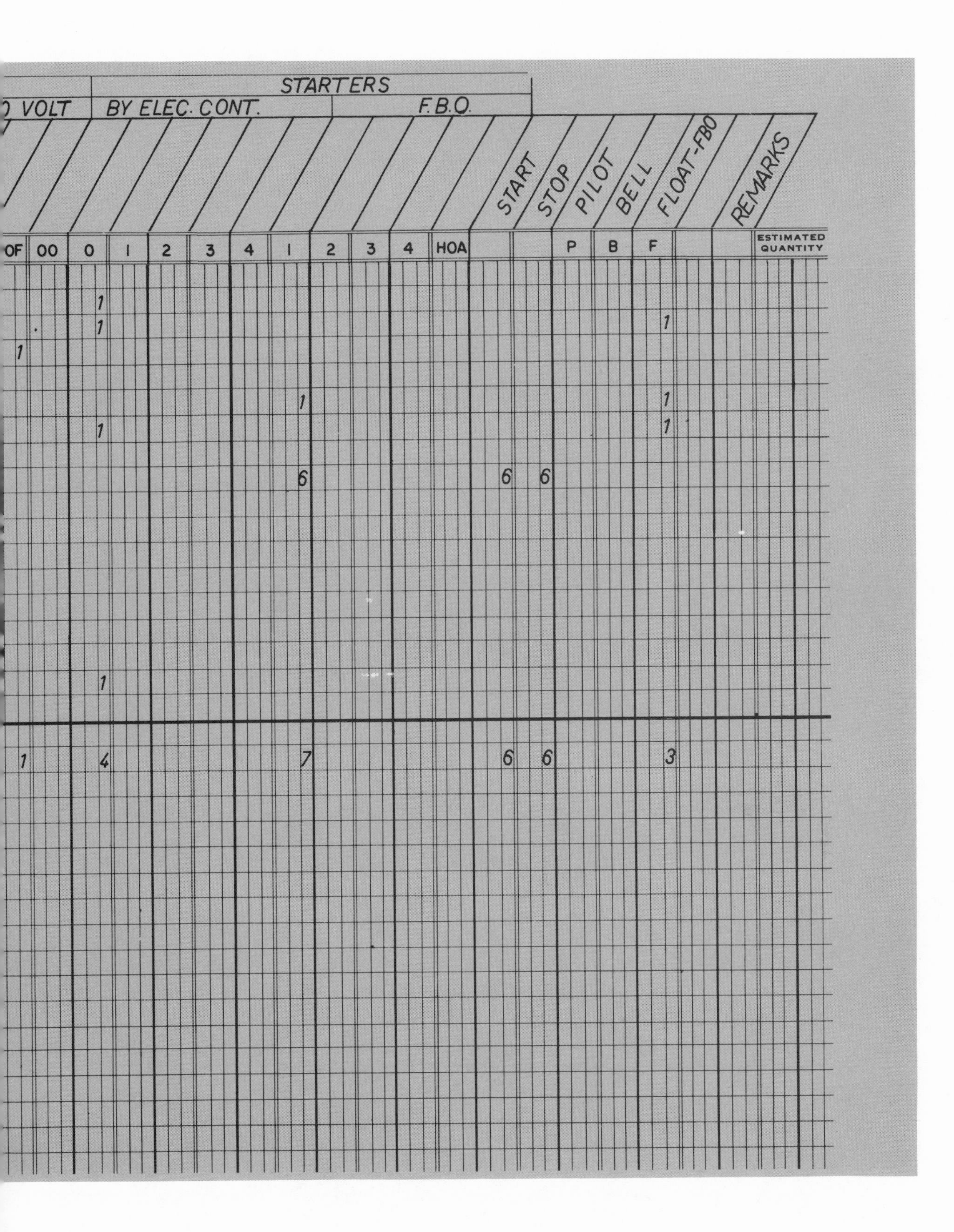

) VOLT		STARTERS																	
		BY ELEC. CONT.					F.B.O.					START	STOP	PILOT	BELL	FLOAT-FBO		REMARKS	
OF	00	0	1	2	3	4	1	2	3	4	HOA			P	B	F			ESTIMATED QUANTITY
		1																	
		1														1			
1																			
							1									1			
		1														1			
							6					6	6						
		1																	
1		4					7					6	6			3			

TABLE TO-5 SERVICE AND POWER RISERS, INCLUDING MOTOR BRANCH CIRCUITS

LOCATION: SERVICE AND POWER (FROM DRAWING E3

ARCHITECT ENGINEER: RISERS, INCLUDES

CLASSIFICATION: MOTOR BRANCH CIRCUITS

	RIGID GALV. COND.								
	3/4	1	1 1/4	1 1/2	2	2 1/2	3	3 1/2	4
FROM XFORMERS TO MS LF									
FROM MS TO HP 3									4/170
FROM MS TO HP 2								6/220	
FROM MS TO HP 1							6/140		
FROM MS TO HV 1								4/120	
FROM MS TO LDP								4/40	
FROM MS TO MDP								4/40	
FROM MDP TO MCCI						15			
FROM MDP TO TRANS.							20		
FROM LVDP TO TRANS.								20	
FROM LVDP TO P5					3/60				
FROM LVDP TO OS CONTACTOR				3/175					
FROM LVDP TO P4					2/90				
FROM LVDP TO PC				4/150					
FROM LVDP TO P3				2/115					
FROM LVDP TO P1					2/200				
FROM LVDP TO P2				2/175					
FROM LVDP TO PB					2/110				
FROM LDP TO L5					3/75				
FROM LDP TO SMP					3/140				
FROM LDP TO LD				2/75					
FROM LDP TO L4					2/85				
FROM LDP TO LC			4/150						
FROM LDP TO L3					2/115				
FROM LDP TO L3					2/115				
FROM LDP TO ELEVATOR					2/130				
FROM LDP TO LB						2/100			
FROM LDP TO LA				2/100					
FROM LDP TO L1				2/200					
FROM HP2 TO HV2B					3/70				
ELBOWS			4	17	26	2	6	18	4
CONDUIT + WIRE TOTALS:									
TABLE TO-5			150	990	1190	115	380	440	170
TABLE TO-6	6200	3070	340						
TABLE TO-7	39500	600	90	300					
TOTAL CONDUIT + WIRE	46000	3700	1600	1300	1200	120	380	440	170

TOR ESTIMATE NO.

SIONS SHEET NO.

XHHW DATE

8	6	4	3	2	1	1/0	2/0	3/0	4/0	MCM 250	300	350	400	450	500		ESTIMATED QUANTITY	UNIT
															800			
												1280						
										1080								
												320						
												280						
												280						
								75										
										90								
												160						
				280														
		780																
				400														
		640																
		500																
				840														
		740																
				480														
				340														
						600												
		340																
				380														
	640																	
				500														
				560														
								440										
440																		
				480														
				340														
440	640	3000		4600		600		515		1170		2320			800			
300	1100	700																
1300	1800	3700		4600		600		515		1170		2320			800			

TABLE TO-6 COMMUNICATION SYSTEM RISERS AND BRANCHES

LOCATION

ARCHITECT ENGINEER SYSTEM RISERS

CLASSIFICATION AND BRANCHES

	RGC			EMT		
	3/4	1	1 1/4	3/4	1	1 1/4
FIRE ALARM SYSTEM	3800		40			
DOOR SECURITY SYSTEM	1600	470	300			
WATCHMAN'S TOUR		2600				
MASTER CLOCK	800					
TOTAL	6200	3070	340			

...TOR	ESTIMATE NO.
...IONS	SHEET NO.
...D	DATE

...ISA	SHIELD TEL.															ESTIMATED QUANTITY	UNIT
50																	
	6000																
50	6000																

TABLE TO-7 BRANCH CIRCUITS

LOCATION

ARCHITECT
ENGINEER

CLASSIFICATION BRANCH CIRCUITS

DESCRIPTION	NO.	DIMENSIONS			RGC				EMT				
					3/4	1	1 1/4	1 1/2	3/4	1	1 1/4	1 1/2	
E2-1					13000	200	90	100					
E2-2					7000					150			
E2-3					15000	400		200					
E2-4					1500				500				30
E2-5					3000								
				TOTAL	39500	600	90	300	500	150			30

TOR

IONS

D

ESTIMATE NO.

SHEET NO.

DATE

XHHW				
8	6	4	ESTIMATED QUANTITY	UNIT
00	500			
	600			
		700		
00	1100	700		

TABLE TO-8 SWITCHBOARDS AND PANELBOARDS (Part 1 of 4)

A	B	C	U	D	E	F	G
	1	MAIN SWBD 277/	EA			94 –	94 –
		480 30 4W w/fusible					
		POWER PROTECTER		Lump sum	48335 –		
		4000 AMPS	EA				
	3	4000 AMP FUSES	EA				
	4	600-3 BREAKERS	EA			10 02	40 —
	2	800-3 BREAKERS	EA			12 12	24 —
	1	PANEL HP1 2 SEC	EA			24 60	25 —
		277 / 400 V 3ϕ 4W	EA				
		800 MLO	EA			18 60	18 —
	1	30-3 SWITCH				96	1 —
	1	40-3 SWITCH	EA			96	1 —
	1	70-3 SWITCH	EA			1 46	1 —
	1	100-3 SWITCH	EA			2 07	2 —
	4	125-3 SWITCH	EA			2 39	10 —
	1	PANEL HPIA	EA			5 02	5 —
		277 / 480 V 3ϕ 4W					
		100 MLO	EA			2 60	3 —
	1	20-1 BREAKER	EA			65	1 —
	10	20-3				96	10 —
	1	PANEL HP2	EA			17 32	17 —
		277 / 480V 3ϕ 4W					
		600A MLO	EA			13 22	13 —
	6	70-3 SWITCHES	EA			1 46	9 —
	2	100-3 SWITCHES	EA			2 07	4 —
	1	PANEL HP2 A	EA			5 02	5 —
		277/480V 3ϕ 4W					
		100 A MLO	EA			2 60	3 —
	4	20-1 BREAKERS	EA			65	1 —
	4	30-3 BREAKERS	EA			96	1 —
		SUBTOTAL					288 —

ABLE TO-8 SWITCHBOARDS AND PANELBOARDS (Part 2 of 4)

A	B	C	U	D	E	F	G
	1	PANEL HP3 2SEC	EA			24 60	25 —
		277/480V 3ϕ 4W		Lump sum		—	
		800 A MLO	EA			18 06	18 —
	2	50-3 SWITCH	EA			1 46	3 —
	1	60-3 SWITCH	EA			1 46	1 —
	1	70-3 SWITCH	EA			1 46	1 —
	1	100-3 SWITCH	EA			1 46	1 —
	1	200-3 SWITCH	EA			3 43	3 —
	1	300-3 SWITCH	EA			5 01	5 —
	1	PANEL LDP 2SEC	EA			17 32	17 —
		277/480V 3ϕ 4W				—	
		600A MLO	EA			13 22	13 —
	1	40-3 SWITCH	EA			96	1 —
	1	50-3 SWITCH	EA			1 46	1 —
	1	70-3 SWITCH	EA			1 46	1 —
	5	100-3 SWITCH	EA			2 07	10 —
	1	150-3 SWITCH	EA			2 62	3 —
	1	200-3 SWITCH	EA			3 09	3 —
	1	PANEL MDP	EA			17 32	17 —
		277/480V 3ϕ 4W				—	
		600A MLO	EA			13 22	13 —
	1	100-3 SWITCH	EA			2 07	2 —
	1	200-3 SWITCH	EA			3 09	3 —
	1	250-3 SWITCH	EA			4 24	4 —
	1	PANEL LVDP	EA			17 32	17 —
		120/208 3ϕ 4W				—	
		600A MLO	EA			13 22	13 —
	5	70-3 BREAKERS	EA			1 46	7 —
	4	100-3 BREAKERS	EA	↓		2 07	8 —
		SUBTOTAL					190 —

TABLE TO-8 SWITCHBOARDS AND PANELBOARDS (Part 3 of 4)

A	B	C	U	D	E	F	G
	1	MOTOR CONTROL MCC1	EA			682	7 —
		480V 3ϕ 3W		Lump sum		—	
		200 A MLO	EA			343	3 —
						—	
	4	15-3 SWITCH	EA			96	4 —
	1	25-3 SWITCH	EA			96	1 —
	1	40-3 SWITCH	EA			96	1 —
	1	60-3 SWITCH	EA			146	1 —
	4	12-2 INTERLOCKS				65	3 —
	1	MOTOR CONTROL MCC2	EA			682	7 —
		480V 3ϕ 3W				—	
		200A MLO	EA			343	3 —
						—	
	5	15-3 SWITCHES	EA			96	5 —
	1	60-3 SWITCHES	EA			146	1 —
	2	12-2 INTERLOCKS	EA			65	1 —
	1	MOTOR CONTROL MCC3	EA			406	4 —
		480V 3ϕ 3W				—	
		100 A MLO	EA			207	2 —
	4	20-3 SWITCHES	EA			96	4 —
	1	20-3 SWITCHES	EA			96	1 —
	2	PANELS SMP - LB	EA			791	16 —
		277/480 3ϕ 4W				—	
	2	225 A MLO	EA			385	8 —
	37	20-1 BREAKERS	EA			65	24 —
	8	20-3 BREAKERS	EA			96	8 —
	20	30-1 BREAKERS	EA	↓		65	13 —
		SUBTOTAL					117 —

BLE TO-8 SWITCHBOARDS AND PANELBOARDS (Part 4 of 4)

A	B	C	U	D	E	F	G
	8	PANELS - L1 - L3 -	EA			406	32 —
		L4 - L5 - E1 -		Lump sum			
		LA - LC - LD					
		277/480 3ϕ 3W					
	8	100 A MLO	EA			207	17 —
	86	20-1 BREAKERS	EA			65	56 —
	8	20-2 BREAKERS	EA			65	5 —
	8	20-3 BREAKERS	EA			96	8 —
	4	30-2 BREAKERS	EA			65	3 —
	3	PANELS PB - PC -	EA			791	24 —
		OS -120-208 3ϕ 4W					
	3	225 A MLO	EA			343	10 —
	103	20-1 BREAKERS	EA			65	67 —
	6	20-2 BREAKERS	EA			65	4 —
	5	20-3 BREAKERS	EA			96	5 —
	4	30-2 BREAKERS	EA			65	3 —
	5	PANELS - P1 - P2 - P3	EA			502	25 —
		P4 - P5 -120/208 3ϕ 4W					
	5	100A MLO	EA			207	10 —
	52	20-1 BREAKERS	EA			65	34 —
	8	20-2 BREAKERS	EA			65	5 —
	4	20-3 BREAKERS	EA			96	4 —
	4	30-1 BREAKERS	EA				
	4	30-2 BREAKERS	EA			65	3 —
	4	40-2 BREAKERS	EA			65	3 —
	1	50-2 BREAKERS	EA	↓		85	1 —
		SUBTOTAL					322 —

TABLE TO-9 FIXTURES AND DEVICES (Part 1 of 2)

A	B	C	U	D		E		F		G	
60/c	310	TYPE A Inc. Lamps	EA	55	—	17050	—	1	69	524	—
C	66	B		28	—	1848	—	1	69	112	—
72/c	409	C		17	—	6953	—	1	39	569	—
C	35	D		9	—	315	—	1	24	43	—
C	17	E		101	—	1717	—	2	50	43	—
B	27	F		38	—	1026	—	2	05	55	—
C	3	G		34	—	102	—	1	60	5	—
C	29	H		29	—	841	—	1	75	51	—
C	1	J		38	—	38	—	1	80	2	—
C	41	K		46	—	1886	—	2	05	84	—
C	6	L		61	—	366	—	1	80	11	—
46/c	274	M		94	—	25756	—	2	25	617	—
C	5	N		105	—	525	—	2	25	11	—
C	4	P		19	—	76	—	1	40	6	—
B	21	EXIT		34	—	714	—	1	75	37	—
C	2	EXP		71	—	142	—	2	—	4	—
B	6	VAULT		12	—	72	—	1	50	9	—
B	2	PIT		18	—	36	—	1	50	3	—
	56	B's	EA		60	34	—	1	—	56	—
	386	C's	EA		60	232	—	1	—	386	—
		DEVICES									
S	70	S 1	EA	2	12	148	—		24	18	—
S	8	S 3	EA	2	70	22	—		35	3	—
S	4	SP	EA	3	05	12	—		34	1	—
S	204	DUPLEX RECEP	EA	2	70	551	—		26	70	—
B	23	CLOCK	EA	1	60	37	—		26	6	—
	23	B's	EA		60	14	—	1	—	23	—
	286	S's	EA		60	172	—	1	—	286	—
		F.A. DEVICES									
B	26	MANUAL STATION	EA	8	20	213	—	1	—	26	—
B	20	HORN w/ LIGHT	EA	50	—	1000	—	1	—	20	—
C	4	FIX TEMP DETECTOR	EA	3	55	14	—		50	2	—
C	85	R-R DETECTOR	EA	3	55	302	—		50	43	—
	1	MASTER BOX	EA	220	—	220	—	4	—	4	—
B	3	SMOKE DETECTORS	EA	62	—	186	—	1	50	5	—
		PAGE TOTAL				62620	—			3135	—

BLE TO-9 FIXTURES AND DEVICES (Part 2 of 2)

A	B	C	U	D		E		F		G	
		F.A. DEVICES cont.									
B	2	MAG DOOR HOLDER	EA	41	—	82	—	1	—	2	—
	1	ANNUNCIATOR 12 Zone	EA	450	—	450	—	6	—	6	—
	1	MASTER CONTROL 12 Zone	EA	640	—	640	—	6	—	6	—
	51	B's	EA		60	31	—	1	—	51	—
	89	C's	EA		60	53	—	1	—	89	—
		WATCHMAN'S TOUR SYSTEM									
B	42	DOOR SWITCH	EA	2	75	116	—	1	—	42	—
B	14	KEY RESET	EA	12	—	168	—	1	—	14	—
B	2	STATION w/ TEL	EA	68	—	136	—	1	50	3	—
B	10	STATION ONLY	EA	17	—	170	—		50	5	—
	68	B's	EA		60	41	—	1	—	41	—
B	18	CABINET HTRS	EA	94	—	1692	—	2	—	36	—
	18	B's	EA		60	11	—	1	—	18	—
	23	CLOCKS	EA	40	—	920	—		50	12	—
		SNOW MELTING MATS									
	28	TYPE A 4.5	sq ft	12	—	336	—	6	75	189	—
	11	C 120	sq ft	120	—	1320	—	18	—	198	—
	5	D 60	sq ft	60	—	300	—	9	—	45	—
	1	F 120	sq ft	140	—	120	—	18	—	18	—
	1	G 120	sq ft	140	—	120	—	18	—	18	—
		PAGE 2 TOTAL				6706	—			793	—
		PAGE 1 TOTAL				62620	—			3135	—
		TOTAL				69326	—			3928	—

TABLE TO-10 MOTOR EQUIPMENT AND HEATERS

A	B	C	U	D	E	F	G
	3	3P-30-208 DISC SW. F	EA	19 —	57 —	2 33	7 —
	4	0 MAG STARTER		38 —	152 —	2 82	6 —
	7	1 MAG STARTER (FBO)				2 94	21 —
	6	START-STOP PB		8 —	48 —	1 99	12 —
	3	FLOAT SWITCHES (FBO)				1 99	6
	4	CONNECT FRAC HP		2 —	8 —	65	3 —
	7	CONNECT 3/4 - 7 1/2		2 —	14 —	97	7 —
	5	CONNECT 10 - 30	↓	2 —	10 —	1 46	7 —
	3	2P-30-208 DISC SW.	EA	16 —	48 —	2 19	7 —
	12	3P-30-480V DISC SW.		33 —	396 —	2 88	35 —
	2	3P-60-480V DISC SW.		40 —	80 —	4 10	8 —
	9	3P-100-480V DISC SW.		74 —	666 —	5 26	47 —
	3	3P-200-480V DISC SW.		107 —	321 —	9 02	27 —
	2	3P-400-480V DISC SW.		277 —	554 —	12 78	26 —
C	12	20 AMP SP IN JB	↓	8 —	96 —	2 36	28 —
	12	C's		60	7 —	1 —	12 —
		HEATERS TO CONNECT	(ONLY)				
	17	1-30 KW	EA	2 —	34 —	97	16 —
	7	35-60 KW		2 —	14 —	1 46	10 —
	4	65-100 KW		2 —	8 —	1 78	7 —
	1	101-185 KW		2 —	2 —	3 73	4 —
	1	186-300 KW	↓	2 —	2 —	6 38	6 —
		PAGE TOTAL			2517 —		302 —

ABLE TO-11 CONDUIT AND WIRE (Part 1 of 2)

B	C	U	D	E	F	G
46000	3/4" RIGID GALV	CF	37 11	17071 —	5 65	2599 —
3700	1		47 10	1743 —	7 45	276 —
1600	1 1/4		61 62	986 —	8 70	139 —
1300	1 1/2		82 81	1077 —	10 30	134 —
1200	2		98 45	1181 —	13 50	162 —
120	2 1/2		152 —	182 —	18 45	22 —
380	3		199 —	756 —	22 55	86 —
440	3 1/2		260 —	1144 —	27 25	120 —
170	4		307 —	522 —	35 70	61 —
500	3/4" EMT	CF	14 52	73 —	4 75	24 —
150	1		20 75	31 —	6 40	10 —
22	4000A CU BUS	LF	182 —	4004 —	3 23	71 —
300	200A PLUG IN BUS DUCT	LF	20 80	6240 —	44	132 —
840	PLUG MOLD STRIP	LF	2 85	2394 —	10	84 —
	WIRE					
59700	14 THHN	MF	49 —	2925 —	6 90	412 —
113600	12	MF	65 47	7437 —	8 00	909 —
1400	10	MF	106 —	148 —	10 40	15 —
1300	8 XHHN	MF	158 13	206 —	11 95	16 —
1800	6	MF	269 —	484 —	13 05	23 —
3700	4	MF	396 —	1465 —	16 40	61 —
4600	2	MF	794 —	3652 —	18 70	86 —
600	1/0	MF	1098 —	659 —	24 68	15 —
515	3/0	MF	1414 —	728 —	30 —	15 —
1170	250 MCM	MF	2103 —	2461 —	37 70	44 —
2320	350 MCM	MF	2774 —	6436 —	43 50	101 —
800	500 MCM	MF	3801 —	3040 —	50 —	40 —
150	1 MSA	MF	245 —	37 —	8 —	2 —
6000	SHIELDED TEL=18 (3)	MF	49 —	294 —	6 90	41 —
	PAGE TOTAL			67376 —		5700 —

TABLE TO-11 CONDUIT AND WIRE (Part 2 of 2)

A	B	C	U	D	E	F	G
		ELBOWS					
	4	1 1/4 R.G.C.	EA	15 26	61 —	45	2 —
	17	1 1/2	EA	20 48	348 —	55	9 —
	26	2	EA	31 47	817 —	90	23 —
	2	2 1/2	EA	58 87	118 —	1 70	3 —
	6	3	EA	94 54	967 —	2 20	13 —
	18	3 1/2	EA	171 65	3090 —	2 75	50 —
	4	4	EA	198 90	796 —	3 50	14 —
		PAGE 2 TOTAL			6197 —		114 —
		PAGE 1 TOTAL			67376 —		5700 —
		TOTAL			73570 —		5814 —

ABLE TO-12 SUMMARY SHEET

ESTIMATE SUMMARY SHEET

OPERATION A SERVICE AND SUPPLY BUILDING — ESTIMATE NO. 130
LOCATION USA — BID DUE 1-2-74
CONST. AND SIZE CAST IN PLACE CONCRETE 87,500SQ. FT. — G.C. BID DUE 1-12-74
ARCHITECT WALLACE FLOYD & ELLEN ZWEIG — COMPLETION TIME 18 MOS.
ENGINEER SHOOSHANIAN ENGINEERS, INC. — ESTIMATE BY L.C.
OWNER — CHECKED BY H.G.
ESTIMATE TO M.B.M.

SHEET NO.	SECTION	DESCRIPTION	MATERIAL		LABOR	
TO-8	1630	SWBDS PANELS	48335		917	HRS
TO-9	1640-1650	FIXTURES & DEVICES (INCLUDING SYSTEM DEVICES)	69326		3928	HRS
TO-10	1670	MOTOR EQUIPMENT	2517		300	HRS
TO-11	1610	CONDUIT & WIRE	73570		5814	HRS
		TOTAL	193751		10961	HRS

SUMMARY

Item	Amount	Notes
MATERIALS	193420	(10,961 x 9.30 PER HR = 101,937 LABOR COST)
MATERIAL SALES TAXES	NONE	
LABOR-PRODUCTIVE	101937	
SUPER. & NON PRODUCTIVE LABOR	32550	
TRAVEL TIME OR EXPENSE	NONE	
INSURANCE & PAYROLL TAXES	44381	
JOB EXPENSES	7446	MAT. & LABOR TOTALS = 372,288
BOND	NONE	
TEMPORARY LIGHT AND POWER 700 HRS at 9.30 PER HR (MATERIALS INCLUDED)	6510	((4.5%)(386,244) = 17,380 ESCALATION COSTS)
ESCALATION 1 1/2 % PER MONTH x 3 = 4.5%	17380	
TOTAL PRIME COST	403624	
OVERHEAD 21%	84761	
TOTAL COST	488385	
PROFIT 10%	48839	
AMOUNT OF ESTIMATE	537224	
PRICE SUBMITTED	520,000	(Price submitted is lower because contractor wants job.)

TEMP. L&P INSTALLATION SEE SUMMARY
TEMP. L&P MAINTENANCE "

ALTERNATES NONE — UNIT PRICES NONE

QUALIFICATIONS: NAT. ELEC. CODE

LABOR RATE 9.30 HR. EXPIRES AUG 75 NEW RATE ______ HR. EXPIRES ______